BETHLEHEM STEEL

BETHLEHEM STEEL

ANDREW GARN

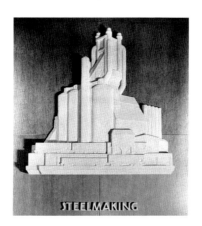

STEELMAKING

PRINCETON ARCHITECTURAL PRESS

Dedicated to Meriwether Garn, George Garn,
Ann Marie Richard, Arthur MacTavish, and Frank Delluva.

Published by
Princeton Architectural Press
37 East 7th Street
New York, NY 10003
212.995.9620

For a free catalog of books, call 1.800.722.6657
Visit our website at www.papress.com

Design: Lisa Mahar
Layout: Adam B. Bohannon
Project editor: Beth Harrison

Special thanks to: Ann Alter, Eugenia Bell, Jan Cigliano, Jane Garvie,
Caroline Green, Clare Jacobson, Mirjana Javornik, Therese Kelly,
Leslie Ann Kent, Mark Lamster, Sara Moss, Anne Nitschke,
Lottchen Shivers, Sara E. Stemen, and Jennifer Thompson
of Princeton Architectural Press
—Kevin C. Lippert, publisher

Library of Congress Cataloging-in-Publication Data
Garn, Andrew.
Bethlehem Steel / photographs, Andrew Garn;
introduction by Lance E. Metz.
p. cm.
Includes bibliographical references (p.).
ISBN: 1-56898-197-X
1. Bethlehem Steel Company. 2. Steel-works—Pennsylvania—
Bethlehem.
I. Metz, Lance E. II. Title.
TS330.B4G37   1999
669'.1'0974822022—dc21   99-23376
                          CIP

This is a Pamphlet Architecture book.

This book was made possible by a grant from Furthermore,
the publication program of The J.M. Kaplan Fund.

Cover photo: Blast furnaces.
Back cover photo: Internal combustion blowing engines in the power house.
Inside front cover: Important products of the Bethlehem Steel
    Corporation.
Title page: Decorative wood carving in lobby of the Bethlehem Steel
    Corporation's Third Street Office Building in Bethlehem,
    Pennsylvania.
Inside back cover: Products produced by the Bethlehem Steel
    Corporation through the 1970s.
Pages viii–ix: Combination rolling mill.

CONTENTS

Acknowledgments

A hearty thank-you goes out to the many people who assisted me in making this book possible. I would especially like to thank Bette Kovach of Bethlehem Steel, who spent countless hours escorting me throughout every crevice of the plant, and Charlie Martin of the Bethlehem engineering department, who provided access to the archives.

I would also like to thank Barbara Hall of the Hagley Museum, Tom Heard of the National Canal Museum, A.M. Richard for her steadfast belief in the project, Ellen Johnson for helping with Arthur, Barbara Flanagan for her shared excitement of the plant, Bennett Beckenstein, and Beth Harrison and Clare Jacobson of Princeton Architectural Press.

Finally, I would like to gratefully acknowledge Furthermore, the publication program of The J.M. Kaplan Fund, which provided a grant to make this book possible.

Foreword

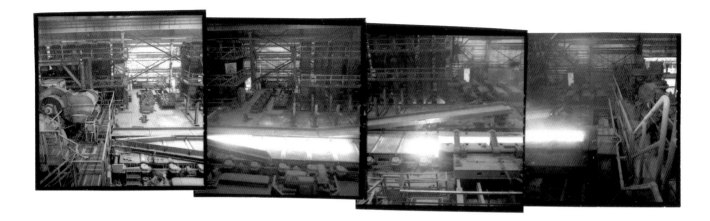

Few historians would argue that one of the most remarkable contributions the United States has made to modern building history of the world is the high-rise skeleton structure. A symbol of twentieth-century economic power, the tallest tower in the landscape is generally taken to mark the center of urban life. Such a landmark often rests not only on several engineering features, but also on standardized elements made of steel. Indeed, for the majority of the twentieth century, the name Bethlehem Steel remained on the minds of most professionals concerned with construction.

The evidence of the company's contribution is in the history books and on the streets around us. When architect Cass Gilbert began his collaboration with Gunvald Aus, Engineers, for the construction of Woolworth Building in Manhattan, 1911–1913, they employed Bethlehem Steel's continuously rolled wide-flange beams. The highest building in the world at the time, this terra cotta–clad cathedral of commerce became a subject for photographers and poets alike. Bethlehem was also involved in the construction of the Merchandise Mart in Chicago, 1930, and architect William Van Alen's Chrysler Building, built the same year in New York, with a stainless steel cap.

Bethlehem steel was also fabricated for some of the country's best-known spans. The classic suspension bridges prepared by their Fabricated Steel Construction Division include the George Washington Bridge, designed by the engineer Othmar Ammann, with Cass Gilbert as architect, which opened in 1931. This 3,500 foot span was nearly double the length of its closest competitor at the time, visually anchored by the imposing space frame steel towers originally to be clad in granite. The Golden Gate Bridge, designed by Joseph Strauss, Othmar Ammann, and others, when completed in 1937 remained the longest span in the country for almost thirty years. It was exceeded only by another Bethlehem span, over the Verrazano Narrows in New York, designed by the engineering firm of Ammon & Whitney and constructed from 1959 to 1964.

After World War II, the glass curtain wall was announced by Skidmore, Owings & Merrill's Lever House, constructed in 1952, again with Bethlehem steel. As is well known, Miesian modernism made its mark with steel. Even the aluminum-clad exterior of the Citicorp Center, by Hugh Stubbins & Associates, with Emory Roth, on Lexington Avenue is carried on Bethlehem steel, as are countless

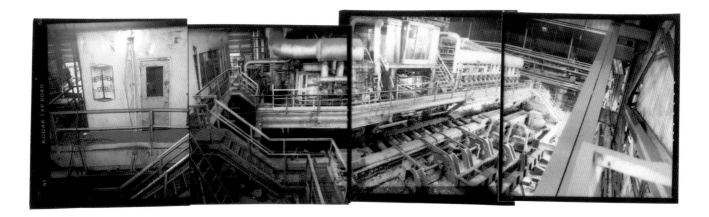

other office buildings, banks, schools, hotels, shopping malls, and parking structures throughout the world.

As historian Lance Metz demonstrates, the growth of the Bethlehem Steel Corporation depended on the railroad and the ability to find new markets. More important than high-rise buildings were the larger military contracts. While in times of peace, the firm might depend upon automobile manufacturing, during war the production of battleship steel, armor plate, and artillery was crucial. Although often in the shadow of the giant, U.S. Steel, Bethlehem still enjoyed the lead in developing several products.

The most obvious display of Bethlehem steel is the plant itself, a true industrial monument. The photography of Andrew Garn provides not only an overview of the site, but also details of the complex mechanical and industrial engineering. It begins to suggest the enormous demand for raw materials—the iron ore, coke, manganese ore, scrap, limestone, gas, and oil—that was continuously being filled. The images today display the industrial forms and spaces, spurring the imagination, suggesting the smoke, smells, noises, and heat that were associated with pouring, rolling, and pounding metal.

Similarly, the text and archival images allude to all the aspirations and pain contained in the same site: the economic problems of the immigrant workers, the racial difficulties, the labor unrest, and the constant management challenges including unions and government. Through the 1950s, steel was generally understood as a fundamental industry, the backbone of a military-industrial complex that allowed the United States and its Allies to win two major wars, and several small conflicts. In the wake of the Vietnam War, cheaper foreign steel was dumped on domestic markets and automobile makers began to feel the pinch of Japanese competition as fuel economy spurred materials substitution. In the postindustrial economy of the 1980s and 1990s, steelmaking seems less important. The glow of the furnaces seen at night from a distance over Bethlehem has faded, as parts of the plant are converted into a museum of our industrial heritage.

Michael A. Tomlan
Associate Professor and Director
Graduate Program in Historic Preservation Planning
Cornell University

BETHLEHEM STEEL

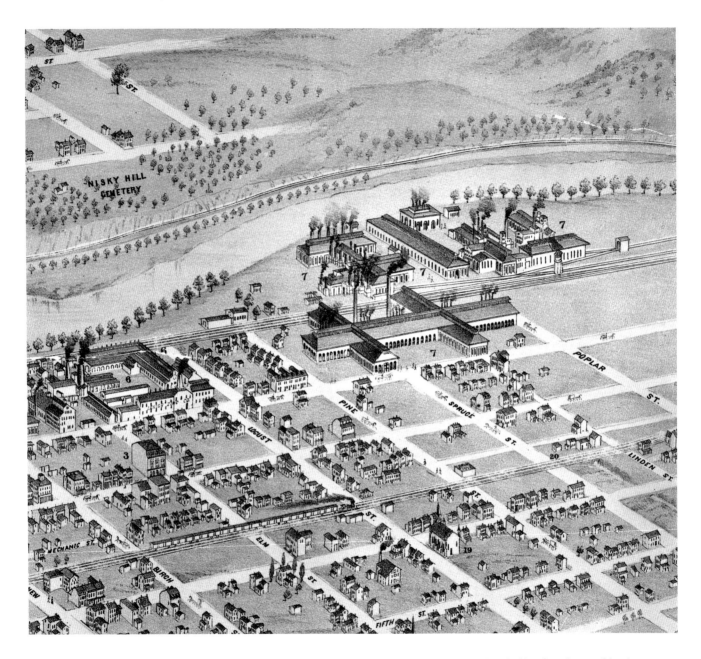

An 1878 bird's-eye view of the Bethlehem Iron Company; the tracks of the Lehigh Valley Railroad divide the steelmaking building from the rest of the plant.

A Short History
of the Bethlehem Steel Corporation

Lance E. Metz

The Beginning:
The Lehigh Coal and Navigation Company

The ultimate origin of the Bethlehem Steel Corporation's plant in Bethlehem, Pennsylvania, can be found in the history of the Lehigh Coal and Navigation Company, which was founded in 1822 by Philadelphia entrepreneurs Josiah White and Erskine Hazard, and which served as the catalyst for the early industrial development of the Lehigh Valley region of eastern Pennsylvania. The Lehigh Navigation waterway system was constructed between 1827 and 1829 by the Lehigh Coal and Navigation Company (LC&N). During this time, a fuel crisis was mounting in the great seaports of Philadelphia and New York. The Lehigh Navigation made it possible for large quantities of anthracite to be shipped to these growing urban centers from the vast coal fields of northeastern Pennsylvania. The anthracite reached these cities via the Delaware and Morris Canals, which connected with the Lehigh Navigation at the city of Easton, Pennsylvania. The fuel crisis, caused by the rapid depletion of the Eastern hardwood forests, was alleviated by this anthracite. Since the colonial era, American iron furnaces had depended on charcoal for fuel. Char-

coal was burned in a blast furnace to heat a mixture of iron ore and limestone to a temperature of 2,800°F, at which point the molten iron was released and the remaining parts of the ore were fused with the limestone to form slag. Depending on its size, an American iron furnace would consume, on a daily basis, charcoal made from a half-acre or more of hardwood forest. Depletion of the forests forced many furnaces east of the Allegheny Mountains to shut down or severely curtail their production during the 1830s.

Experiments in America using anthracite as a substitute blast-furnace fuel had been conducted in many places, none of them successful until 1840 because no one fully understood the hot blast process before then. In that year, Welsh ironmaster David Thomas, using capital, land, and resources provided by the LC&N, placed in operation the first furnace of the Lehigh Crane Iron Company at what is now Catasauqua, Pennsylvania. Once it became evident that Thomas had overcome the primary technological obstacles to the use of anthracite as a blast-

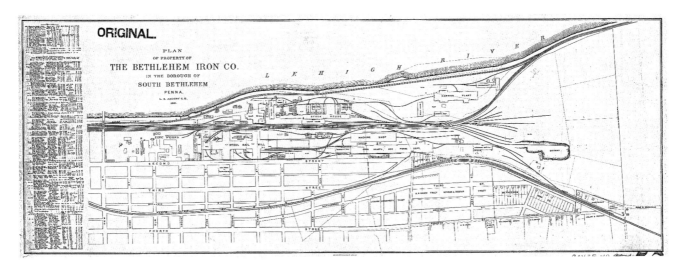

PLAN
OF PROPERTY OF
THE BETHLEHEM IRON CO.
IN THE BOROUGH OF
SOUTH BETHLEHEM
PENNA.

furnace fuel, the Lehigh Valley soon became the center of the American iron industry.

Interacting geological and geographical factors greatly facilitated the rapid growth of the Lehigh Valley as a center of iron production. The LC&N could deliver almost unlimited supplies of anthracite at cheap prices; it could also be used as a source of waterpower to drive blowing engines and other production machinery. The Lehigh Valley region contained large deposits of both moderate-grade iron ore and limestone suitable for use as furnace flux. High-grade iron ore from nearby northcentral New Jersey was transported on the Morris Canal to the Lehigh Valley, where it was mixed with local ores and used in blast furnaces to produce a superior grade of cast iron. The LC&N and its connections with the Delaware and Morris Canals gave Lehigh Valley ironmasters an economical and efficient means of sending their products—principally pig iron, a crude iron that is refined to produce steel, wrought iron, or ingot iron—to the major markets of Philadelphia and New York. All of these factors combined to attract many entrepreneurs who were eager to invest in Lehigh Valley ironworks.

In 1873, 46 percent of all iron produced in the United States came from anthracite furnaces, and 72 percent of those furnaces were in Pennsylvania. The Lehigh Valley region contained the greatest concentration of anthracite-fueled iron furnaces in Pennsylvania, with forty-seven blast furnaces at eighteen locations. Among these was the plant of the Bethlehem Iron Company, located in what was then South Bethlehem. (Bethlehem, South Bethlehem, West Bethlehem, and Northampton Heights all merged to form the present city of Bethlehem in 1919. Today the former South Bethlehem is referred to as the South Side.)

The Railroad Connection

Formed in 1860, the Bethlehem Iron Company, unlike the other Lehigh Valley iron furnace complexes, was not established to produce merchant pig iron, the basic product of the American iron industry for much of the nineteenth century. (Merchant pig iron producers did not cast products themselves, but made cast-iron bars that were sold to foundries and rolling mills. These concerns remelted the bars, or pigs, to produce a vast array of iron items from household hardware to industrial machinery.) The company was founded, instead, to produce iron rails. When other iron companies failed, the Bethlehem Iron Company continued to develop and prosper, largely as a result of the key role that the Lehigh Valley Railroad played in the former company's founding and development.

Left: Plan of the Bethlehem Iron Company, 1891.

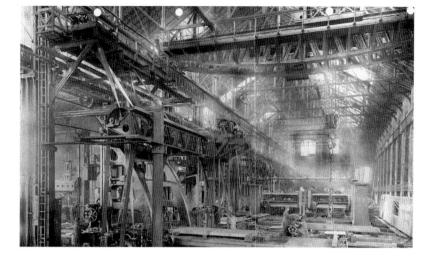

No. 3 Machine Shop c. 1900; this shop specialized in the machining and fitting of armor plate.

The Lehigh Valley Railroad was the creation of a former LC&N canalboat captain and boat builder, Asa Packer (1805–1879). Packer had made a fortune leasing and mining LC&N anthracite deposits and by speculating in real estate in the rapidly growing mining towns of what are now Carbon and Schuylkill Counties. In 1851 he purchased control of the Delaware, Lehigh, Schuylkill, and Susquehanna Railroad, which he eventually reorganized into the Lehigh Valley Railroad. Chartered but as of yet unbuilt, this railroad would provide Packer with the means to transport anthracite and to challenge the LC&N's virtual monopoly of the region's commerce. Rail transport was faster and more flexible than canals, but not necessarily cheaper. Railroads could, however, ship freight during the entire year, while canals were frozen during the winter months and could only be built where there was a regular water supply.

Among the most important of Packer's early actions as the controlling stockholder of the railroad was the appointment of Robert H. Sayre (1824–1907) to be the railroad's chief engineer and supervise its construction. Like Packer, Sayre was a former employee of the LC&N, where he had acquired considerable experience as the chief assistant to the company's brilliant chief engineer, Edwin A. Douglas. Under Sayre's direction, a route that paralleled the LC&N

between the anthracite mining area and Easton was quickly surveyed. By the time Packer's corporation was officially reorganized as the Lehigh Valley Railroad, construction was well underway and on June 11, 1855, the railroad was operational. It became a profitable enterprise and a serious competitor to the LC&N almost immediately.

During the course of the Lehigh Valley Railroad's construction, large quantities of rails were purchased from the Lackawanna Iron and Coal Company, located at what is now Scranton, Pennsylvania. It was one of the largest American manufacturers of rails, but the rails were of poor quality. The company was controlled by Moses Taylor, a New York entrepreneur who was also the chief investor in the Delaware, Lackawanna, and Western Railroad. This railroad was, at that time, being built to haul anthracite from northeastern Pennsylvania to connecting railroads in New Jersey. Thus, every purchase of Lackawanna rails by the Lehigh Valley Railroad was in effect serving as a subsidy for a potential competitor. The management of the Lehigh Valley Railroad had few options since all domestic rail manufacturers produced, at best, a product of mediocre quality. Superior British rails from the Bessemer Company were expensive due to high tariffs. The solution to this dilemma was eventually provided by Robert Sayre.

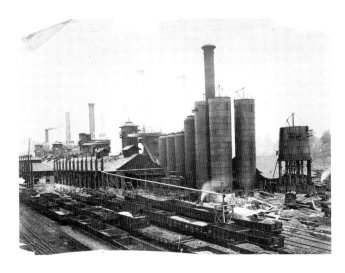

Construction of the E blast furnaces, 1906.

Sayre retained his position as chief engineer after the completion of the Lehigh Valley Railroad and was given additional responsibilities as the line's general superintendent. His authority and independence of action were increased by Asa Packer's tendency to serve as an absentee owner who devoted the majority of his time to dealing and socializing with financiers in Philadelphia and New York. Under Sayre's astute management, the Lehigh Valley Railroad rapidly increased its traffic and by 1858 Sayre had permanently located its general headquarters at the new community of South Bethlehem. His choice of locale was based on that community's proximity to the railroad's primary repair shops and its junctions with the North Pennsylvania Railroad and the Central Railroad of New Jersey, which provided outlets to Philadelphia and New York, respectively. Sayre's move to South Bethlehem also placed him at the geographical center of the Lehigh Valley's rapidly developing iron industry.

In 1857, a Bethlehem merchant, Augustus Wolle, became interested in the development of local iron ore deposits and organized the Sauconna Iron Company. Among the initial subscribers was Asa Packer, who directed Sayre to take an active role in the Sauconna Iron Company's affairs. Realizing that the nascent enterprise could provide a solution to his railroad's difficulties in obtaining rails, Sayre used the financial resources of the Lehigh Valley Railroad to take effective management control of the Sauconna Iron Company. In 1858 he reorganized it as the Bethlehem Rolling Mill and Iron Company, a name that better reflected its intended purpose. Under Sayre's direction, the plant was built along the Lehigh River at a site adjacent to the junction of the Lehigh Valley and North Pennsylvania railroads. This choice of site meant that products could be readily shipped to market in Philadelphia, New York, and the anthracite mining regions of Pennsylvania. Sayre also selected the skilled ironmaster who would be needed to design the plant and supervise its operations: at the inaugural meeting of the board of the Bethlehem Rolling Mill and Iron Company, the directors hired John Fritz to be their general manager and superintendent.

John Fritz and the Three-High Mill

John Fritz (1822–1913) was the most mechanically innovative of America's nineteenth-century ironmasters. He had served since 1854 as the superintendent of the Cambria Iron Company at Johnstown, Pennsylvania. In 1857 he had developed an ingenious "three-high" rail mill that made it

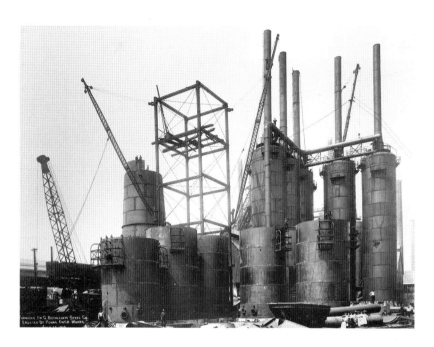

Construction of hot blast furnaces F and G, 1910.

possible, for the first time in America, to produce wrought-iron rails of uniformly high quality at an economical price. Unlike the commonly used two-high rail mill, so called because it had two sets of rolls, the three-high mill enabled a heated iron bloom to be completely rolled into a finished rail before it could cool and potentially shatter. Fritz was granted a patent on his three-high rail mill on October 5, 1858. Two years later, he accepted Sayre's offer and moved to South Bethlehem, arriving on July 5, 1860.

On July 16, 1860, ground was broken for the industrial complex that would eventually evolve into the Bethlehem plant of the Bethlehem Steel Corporation. The general layout of the plant was selected by Robert H. Sayre, while the actual design of the production machinery and the buildings that would contain it was the work of John Fritz. Fritz had experienced several major fires at Cambria, where the buildings were made of wood, and was determined that the Bethlehem Rolling Mill and Iron Company's buildings would be constructed of more durable and fire-resistant materials.

A gifted amateur architect who possessed a passion for combining functionality and aesthetics, Fritz chose local stone as his primary building material since it was both durable and readily available. The roofs of the plant's buildings were supported by an intricate series of wrought-iron trusses and covered with Lehigh Valley slate. To provide adequate natural lighting, Fritz used simple windows of a generally Romanesque configuration; to gather additional natural light, he used high monitors on the center lines of the roofs. These massive buildings covered a rolling mill complex with eight double-puddling furnaces (for the conversion of cast iron into wrought-iron for rail production), six heating furnaces, and an improved three-high rail mill.

The company was reorganized early in 1863 as the Bethlehem Iron Company. On July 27 of that year, the puddling furnaces began producing wrought-iron blooms for the rolling mill, which began to produce rails in September. The rails quickly gained a reputation for excellence and they had a ready market: the parent company, Lehigh Valley Railroad,

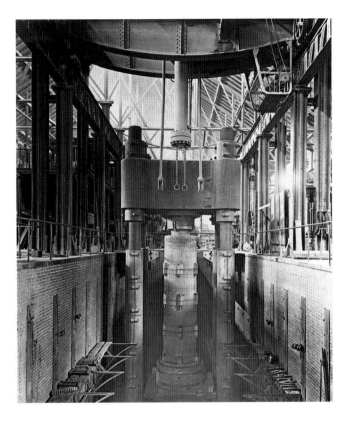

Whitworth fluid compression press, c. 1894. Molten steel was poured into a mold and subjected to hydraulic pressure until the ingot cooled.

purchased almost all of the initial production. The Central Railroad of New Jersey, whose president, John Taylor Johnston, was a member of the Bethlehem Iron Company's board of directors, also became a major customer. By the end of 1863, the iron works had grown to include four stationary steam engines, one blast furnace, fourteen puddling furnaces, nine heating furnaces, a 21-inch (based on the diameter of the rolls) puddle train, and a 21-inch rail train.

The company's second blast furnace was completed in 1867. In his autobiography, John Fritz referred to it as being of an unusual "crinoline" design because of its resemblance to fashionable ladies' skirts of that time. Instead of being constructed of boiler plate, it was made of bands, 7/8" thick by 8" wide, riveted to 8"-in-diameter wrought-iron uprights on 30-inch centers. The entire structure was

lined with refractory material. It had small viewing ports at various places to allow the operators to inspect the furnace for malfunctions. During this same period, John Fritz also supervised the construction of a large machine shop and foundry complex, the products of which greatly expanded the company's market. To further increase its pig-iron production, the company purchased an unused blast furnace on an adjacent property from the Northampton Iron Company, raising annual ironmaking capacity to 30,000 tons.

The high quality of its product soon won a major share of the eastern railroad-rail market for the Bethlehem Iron Company. However, when British Bessemer steel rails began to appear in America during the 1860s, their superior durability attracted the attention of many railroads.

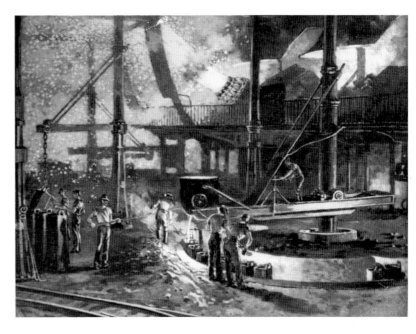

Operating the Bessemer converter, c. 1880s.

Although far costlier than wrought-iron rails, the steel rails lasted six times longer. As early as 1864, the Lehigh Valley Railroad, under Sayre's direction, began to import Bessemer steel rails. The LC&N at that time was extending its Lehigh and Susquehanna Railroad to parallel almost the entire route of the Lehigh Valley Railroad, using the imported steel rails. (The initial shipment was less than a dozen rails, an experiment to text their durability. Sayre did not want to pay the high tariffs that would be entailed in bringing in enough rails to retrack the entire line.) But he also feared that the imported rails would greatly reduce the Lehigh and Susquehanna's maintenance costs, thus giving it an economic advantage over the Lehigh Valley Railroad.

The Introduction of Bessemer Steelmaking

Sayre began to prod the Bethlehem Iron Company to investigate the production of Bessemer steel rails. Early experimentation with new steelmaking processes had been less than successful, and Fritz stood in opposition to the company entering into this unproven area. Years of experimentation produced results, however, and the Pneumatic Steel Association was created in 1865. This was a cartel founded by Alexander Holley, the British engineer who had brought knowledge of the Bessemer process to America; Daniel K. Morrell, general manager of Wood Morrell and Company, the operators of Johnstown's Cambria Iron Company; and Holley's employers, ironmasters John Griswold and John F. Winslow of Troy, New York. The cartel acquired the right to use several steelmaking patents. The Bethlehem Iron Company became a member of the organization in 1867.

The entry of Bethlehem into the Pneumatic Steel Association propelled John Fritz to the forefront of the efforts to create a viable Bessemer steel industry in the United States. He played a large role in the design of the works of the Pennsylvania Steel Company at what is now Steelton, Pennsylvania, the first commercially successful Bessemer steel plant in America. In 1868, Fritz went to Europe to examine steel works in England, France, Germany, and Austria. When he returned from his trip he began to work

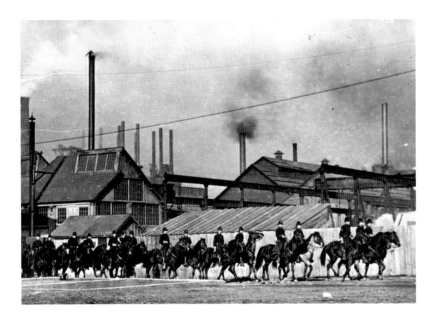

State police patrolling the Poplar Street gate of the plant during a 1910 workers' strike.

with Alexander Holley. Because of Fritz's desire to make Bethlehem's plant the most mechanically efficient of the Bessemer steel works in the United States, it was not placed in full operation until October 4, 1873.

Designing the Steel Plant

In many ways, the Bessemer steel plant that Fritz designed for the Bethlehem Iron Company can be considered the first serious attempt to integrate the production of steel and of iron rails, an achievement recognized early on by Fritz's contemporaries. Robert W. Hunt, a pioneering metallurgist, chemist, and mechanical engineer who was involved in some of the earliest attempts to create a Bessemer steel plant in America, described Fritz's plant in the following passage from his work "A History of Bessemer Manufacture in America," which appeared in Vol. 5 (1876–1877) of the *Journal of the American Institute of Mining Engineers*:

> He arranged his melting-house, engine room, converting-room, blooming and rail mills, all in one grand building, under one roof, and without any partition walls. He placed his cupolas on the ground and hoisted the melted iron on a hydraulic lift and then poured it into the converters. The spiegel is also hoisted and poured into the vessels.... Instead of depending upon friction to drive the rollers of the tables, Mr. Fritz put in a pair of small reversing engines.

A more complete description of the plant and its operations is provided in the following passage from the 1873 *Guide Book of the Lehigh Valley Railroad*:

> The largest manufacturing establishment here is that of the Bethlehem Iron Company, including within its operations, which began in January, 1863, furnaces, rolling mills, machine shop and foundry. Its capital stock is $1,000,000. The measurement of the three stacks is as follows: No. 1, 15 by 63 feet; No. 2, 15 by 45 feet; No. 3, 14 by 50 feet. Their combined capacity is about 30,000 tons per annum. The largest part is used in the adjoining rolling mill, whose capacity is 20,000 tons per

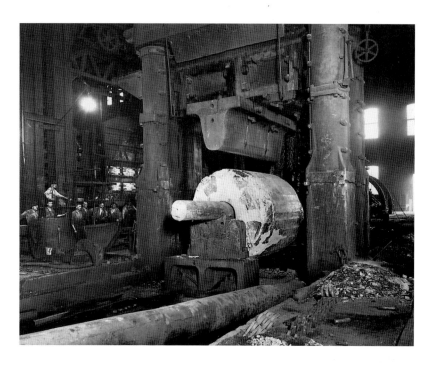

Forging a
battleship
gun, c. 1900.

annum. Its consumption of raw materials is 70,000 tons of Pennsylvania hematite and New Jersey magnetic ore and from 70,000 to 75,000 tons of coal. The total number of men employed at the works proper is about 700. The new building now erecting for the manufacture of iron and steel will be, it is said, the largest in this country and one of the largest in existence anywhere. It will be 105 feet wide spanned by an iron and slate roof without supporting columns. It is 30 feet high to the eaves and is in the shape of a double cross of which the long arm (or main building) is 941 feet and the short arms 140 feet each, making the area covered 1,493 by 105 feet. This is only surpassed by the mill at Creuzot [Le Creusot] in France, which consists of three buildings 60 by 1,400 feet each.

Fritz sought to reduce costs by fabricating almost all of the machinery needed to produce Bessemer steel at the Bethlehem plant. Despite this decision, the uniqueness of his overall design insured that the construction costs of the steel mill would be very high. Fritz liked to take his time designing and fabricating machinery. Alexander Holley once suggested to Fritz that he might have built certain machinery to be unnecessarily strong; Fritz laughed and stated, "Well, if I have, it will never be found out" (Fritz, *The Autobiography of John Fritz*).

Because Fritz was an empirical engineer who arrived at his final design through experimentation and tinkering, he developed an important reconception of American Bessemer steelmaking practice. The entire Bethlehem Bessemer works—converters, ingot reheating furnaces, a three-high blooming train and three-high rail mill—were installed in a single monumental structure. Resembling a 931-foot-long basilica with a 111-foot clear-span nave and two 386-foot-long transepts, this grand building was constructed of local South Mountain stone with arched side openings. Wrought-iron trusswork supported a slate roof with a continuous monitor.

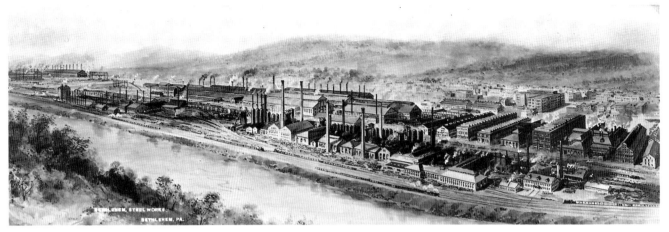

Panoramic view of the Bethlehem plant, 1907.

Fritz's entire design was predicated on the separation of the major operations from each other in order to obviate the problems of intolerable noise and heat concentration that had plagued earlier American Bessemer steel mills. The Bessemer mill occupied about 50 percent of the building's west end and the west crossing of the nave. Fritz placed the blowing engines, which provided blast to the converters, in the transept north of the crossing, leaving room for future expansion in the south transept and in the space adjacent to the Bessemer mill. The cupolas, in which pig iron from the blast furnaces was remelted, were located about 75 feet behind the converters on a platform raised above the general level to the minimum height needed to allow a locomotive to pass in front of and between the converter vessels. Ground transportation was necessary because the distance between the cupola and the vessels was too short and the elevation differential too shallow to allow the use of gravity iron runners to move hot metal, such as had been pioneered by Alexander Holley. (Gravity runners were iron troughs that allowed pig iron to flow by gravity from the cupola furnaces to the Bessemer converters.) Minimal elevation of the cupolas made it possible to accommodate the melting department under the 29-foot clearance below the iron roof trusses.

Despite its innovative design, the Bessemer building had several potential production disadvantages that were caused by the dispersion of the major components and the transportation of pig iron on the ground level. Fritz's design involved additional materials handling, which consumed precious time. Another added expense was that cupola slag had to be removed by rail cars instead of simply being discharged by means of a gravity chute, an iron trough used to remove slag, the waste produced during iron- and steelmaking. Spreading out the production machinery also consumed more real estate, land that could have been used for some other plant purpose. Fritz's design approach also entailed increased building costs.

The building that housed the Bessemer plant was itself little more than a protective shed. None of the equipment essential to Bessemer steel was structurally tied to the building's walls. With the possible exception of the top-supported cranes, the production process would not have collapsed if the building had been removed. The building is insistent in its regularity and its highly autonomous plan; the result is a hybrid that blended a form evolved for the craft-based production of wrought iron with the rational layout required for the high-volume production of the Bessemer process.

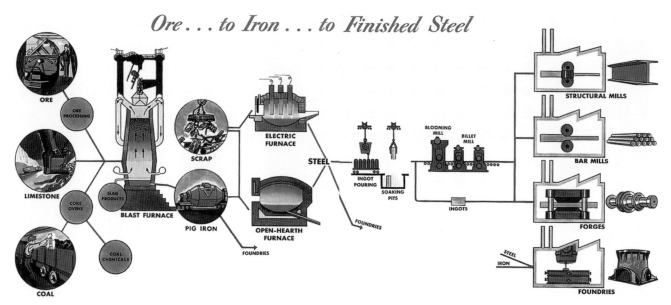

Ore . . . to Iron . . . to Finished Steel

The steelmaking process from ore to iron to finished steel.

Steel Production Begins at Bethlehem

Due to its corporate links to the Lehigh Valley Railroad, only the Bethlehem Iron—of all of the iron furnace complexes in the Lehigh Valley—had the capital resources to undertake steel production. It was thus largely immune to the deleterious effects of eventual declines in the market for merchant pig iron, which so drastically affected the profits of the other iron companies in the region, and it was much better positioned to survive the depression that was brought about by the financial panic of 1873. During this period, other area iron manufacturing facilities either temporarily ceased production or were forced into bankruptcy, and the Lehigh Valley began to cede its preeminence as the leading iron-producing region of the United States to plants in the Pittsburgh region, especially those owned by Andrew Carnegie. On October 4, 1873, the Bessemer converters at the Bethlehem Iron Company were first put into operation and on October 18 the first steel rails were rolled there. Within a year, Bethlehem's new plant had established a national reputation for its productivity and for the quality of its rails.

Soon after the successful completion of the Bessemer steel works at Bethlehem, the company also increased its pig-iron production capacity because, although the demand for merchant pig iron was falling, the demand for rails was rising, and hence more pig iron was needed to produce the steel rails. In 1874–1875 two new furnaces were built, Nos. 4 and 5. These furnaces were 70 feet high and had diameters of 17 feet at the bosh (the bottom or floor of the furnaces; also the area where molten iron collects before it is tapped), larger than almost all other American blast furnaces in use at that time. These engines were blown by extremely powerful horizontal blowing engines. With demand for pig iron still growing, No. 6 Furnace was built in 1881. It was 70 feet in height and 17 feet in diameter at the bosh. Around the same time, the company acquired the North Penn Furnace in the nearby town of Bingen, Pennsylvania, and designated it as No. 7. For a brief period in the 1890s, the company also leased and operated the Lucy Furnace near Dam No. 8 (the Chain Dam) of the Lehigh Navigation near Glendon.

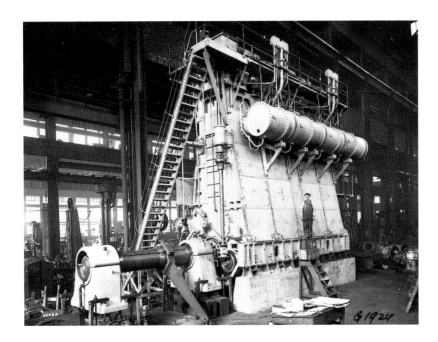

Steamship engine
constructed at the
Bethlehem plant,
c. 1919.

The Military-Industrial Complex Emerges

Although the company continued to prosper during the early 1880s, its share of the American railroad-rail market began to decline as a result of increasing competition from growing Pittsburgh-based firms such as the Carnegie Steel Company. Clearly, if the company was going was to remain viable over the long term, a new product line had to be developed. Therefore, once the U.S. Navy decided to rebuild and modernize its fleet, Bethlehem built a heavy-forging plant. This new demand for a modern American navy of steam-driven steel warships soon began to shape the course of the Bethlehem Iron Company's future development.

Although the United States had been among the world's strongest and most innovative maritime forces during the Civil War, the end of hostilities had brought about rapid naval disarmament. National energies were deflected toward settlement of the West and the rebuilding of the war-ravaged South. The nation's ironclads, steam cruisers, and gunboats were mostly sold abroad or tied up

to rot in the generally inactive navy yards. Almost no new ordnance was produced, and new technology was neglected. By 1881 a series of embarrassing international incidents highlighted the deplorable condition of the U.S. fleet. The growing perception that a strong navy was needed to protect U.S. trade and prestige made possible the beginnings of what would become a sustained effort to create a modern battle squadron.

In 1883, Secretary of the Navy William E. Chandler and Secretary of the Army Robert Todd Lincoln appointed Lt. William Jaques to the Gun Foundry Board. Jaques was sent on several fact-finding tours of European armament makers and on one of these trips he formed business ties with the firm of Joseph Whitworth of Manchester, England. He returned to America as Whitworth's agent and, in 1885, was granted an extended furlough to pursue this personal interest. This type of activity, where government employees become linked to private concerns, marked the beginning of what would become known as the military-industrial complex.

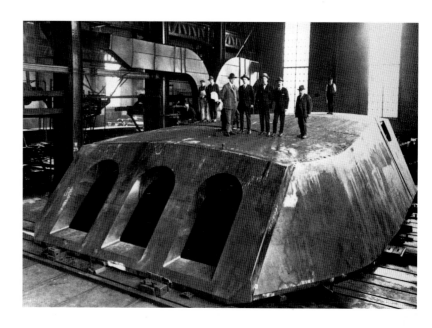

Company officials standing on the assembled main gun turret of the battleship USS *Pennsylvania*.

Jaques was aware that the U.S. Navy would soon solicit bids for the production of heavy guns and other products such as armor that would be needed to further expand the fleet. Jaques contacted the Bethlehem Iron Company with a proposal to serve as an intermediary between it and the Whitworth Company, so that Bethlehem could erect a heavy-forging plant capable of producing ordnance for the U.S. Navy. In 1885, John Fritz, accompanied by Bethlehem Iron Company directors Robert H. Sayre, E.P. Wilbur, William Thurston, and Joseph Wharton, met with Jaques in Philadelphia, where they discussed the feasibility of Whitworth's supplying the technology that Bethlehem needed. In early 1886, a contract between Bethlehem Iron and the Whitworth Company had been executed.

During the spring of 1886, Congress passed a naval appropriations bill that authorized the construction of two armored second-class battleships, one protected cruiser, one first-class torpedo boat, and the complete rebuilding and modernization of two Civil War–era monitors. The two second-class battleships (the USS *Texas* and the USS *Maine*) would have both large-caliber guns (12" and 10"

respectively) and heavy armor plate. The navy solicited bids on August 21, 1886, for 1,310 tons of semi-finished gun forgings and 4,500 tons of steel armor plate. Bethlehem successfully secured both the forgings and armor contracts on June 28 of the following year.

Between 1888 and 1892 the Bethlehem Iron Company completed the first heavy-forging plant to be built by an American steel company. It was designed by John Fritz with the able assistance of Russell Wheeler Davenport (1849–1904), who had entered Bethlehem's employ in 1888. By the autumn of 1890, Bethlehem Iron was successfully delivering gun forgings to the U.S. Navy and was thus able to devote its energies to the completion of the facilities that would be necessary to provide armor plate.

Open-Hearth Steelmaking and Armor Plate Production

This military effort began with the introduction of open-hearth steelmaking technology to the United States. Open-hearth furnaces gave steelmakers the ability to control their raw materials more precisely, thus enabling them to produce ingots possessing exactly the physical and chemi-

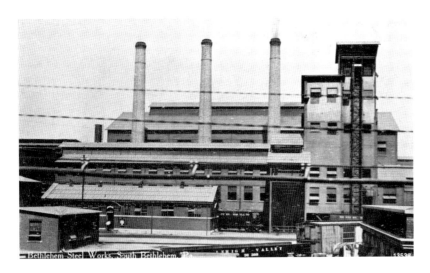

No. 3 Open
Hearth Shop,
c. 1910.

cal properties that were desired. These furnaces made steel of higher quality and greater strength than the more commonly used Bessemer steels. Open-hearth technology was introduced into America by Abram S. Hewitt, the principal proprietor of the firm of Cooper and Hewitt, which operated a large iron and steel works at Trenton, New Jersey. The spread of the new technology was not rapid; by 1880 less than 10 percent of American steel was produced by this method. However, the growing use of open-hearth steel in applications such as bridge building, where a high tensile strength was required, brought about its employment in shipbuilding. By 1880, American shipbuilders and steelmakers had developed a body of useful experience in the utilization of open-hearth steel. Of great significance, open-hearth steel became the preferred material for the production of ordnance armor plate and propulsion machinery parts.

An open-hearth steel furnace is an enclosed rectangular brick structure containing a depressed elongated saucer-shaped floor or hearth. The hearth is charged with a mixture of scrap steel and molten pig iron; the charge is then swept by tongues of flames from burners at each end until the temperature is raised to 3,000°F. The flames are produced in two large chambers containing a checkerwork arrangement of fire bricks through which air or gas can flow freely. These chambers, known as regenerators, are located at opposite ends of the furnace, below the level of the hearth. Each of the regenerators is heated alternately by the products of the furnace's combustion. When one regenerator has exhausted its supply of heat to the open hearth, the direction of the air flow into the furnace is reversed by valves, so that the hot chamber at the opposite end of the open hearth becomes the source of heat flow to melt the charge. The cool chamber is reheated by absorbing the high temperatures produced by the furnace's combustion process.

Despite the mechanical genius of John Fritz and the metallurgical knowledge of Russell Davenport, the Bethlehem Iron Company experienced great difficulties in commencing production of armor plate. And Fritz and Davenport built upon the best contemporary European forging technology to produce a plant of unprecedented size and capacity, these difficulties were not easily overcome. The complexity arises from the fact that armor plate must be extremely hard on the outside in order to break up shells that are fired at it, yet have a supple backing to absorb the impact of those shells.

By 1890, it became clear to the navy that Bethlehem

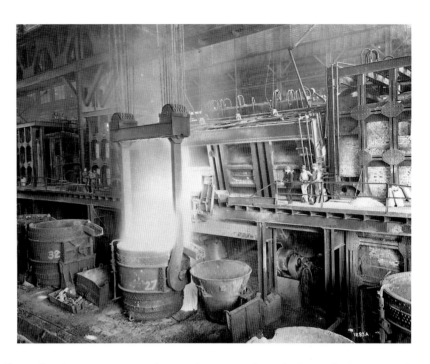

Tapping the open hearth furnace.

would not be able to meet its contract deadlines for the delivery of armor plate. This delay greatly hindered the completion of such major warships as the first modern American battleships, the USS *Maine* and the USS *Texas*. Secretary of the Navy Benjamin Franklin Tracy began negotiations with the Carnegie-Phipps Company of Pittsburgh for the construction of an armor mill and on November 20, 1890, the navy signed a contract with Carnegie-Phipps for 6,000 tons of armor plate. Significantly, the contract specified that the plate could be manufactured from simple steel or nickel steel.

During the late 1880s, French steelmakers had conducted experiments with nickel-alloy steel armor plate, demonstrating a marked superiority of this product over other steel armors. As early as the autumn of 1889, the U.S. Navy secured an experimental batch of Schneider nickel-steel armor. A series of comparative tests were held during September 1890 to evaluate the protective strength of nickel-steel plates, simple steel plates, and a British compound wrought-iron and steel plate. The compound plates were quickly demolished under bombardment by 6-inch and 8-inch armor-piercing projectiles, while the nickel-steel plate appeared to exhibit only marginally better resistance to penetration than did the simple steel plate. However, nickel-steel plate was much more resistant to cracking, which was the greatest fear of the navy's technical experts.

Following the results of the tests, on September 17, 1890, the navy secured an emergency $1,000,000 congressional appropriation for the purchase of nickel. The ultimate effect of these tests was to induce the navy to specify nickel steel as the basis for all orders of armor plate. Since Joseph Wharton, a director and major stockholder of the Bethlehem Iron Company, also possessed considerable holdings in the nickel ore business, Bethlehem was readily able to produce nickel steel.

The production of armor plate was further revolutionized during the early 1890s by the introduction of the

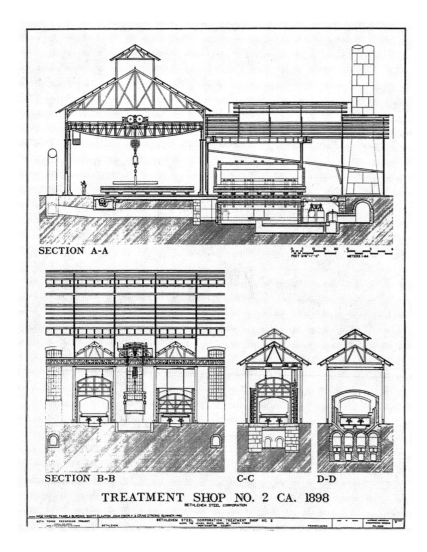

SECTION A-A

SECTION B-B   C-C   D-D

TREATMENT SHOP NO. 2 CA. 1898
BETHLEHEM STEEL CORPORATION

No. 2 Treatment
Shop, c. 1898.

Harvey, or face-hardening, process. This process involved covering low-carbon steel with charcoal and heating it in furnaces, which imparted to the surface of the metal a tough veneer of extraordinary hardness. A variation of the centuries-old cementation process, it was the invention of an American industrialist, Hayward Augustus Harvey, who since 1885 had managed a small plant in Jersey City, New Jersey, which produced specialty hardened steels for the production of tools such as files, cutlery and axes. Harvey's successful enterprise was recapitalized in 1888 and moved to a larger factory at Brill's Station in Newark, New Jersey.

By March 1891 a tentative agreement had been negotiated between the U.S. Navy and the Harvey Steel Company

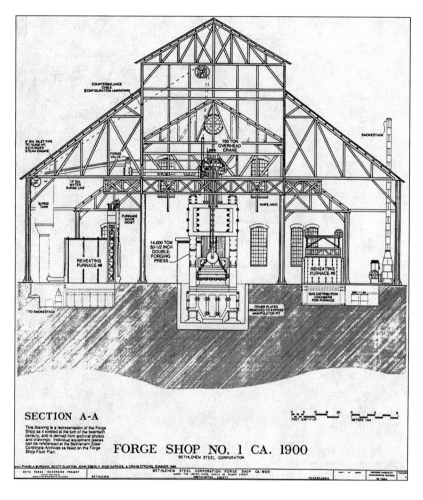

Cross-section of
No. 1 Forge Shop,
c. 1900.

SECTION A-A

This drawing is a representation of the Forge Shop as it existed at the turn of the twentieth century, and is derived from archival photos and drawings. Individual equipment pieces can be referenced at the Bethlehem Steel Corporate Archives as listed on the Forge Shop Floor Plan.

FORGE SHOP NO. 1 CA. 1900
BETHLEHEM STEEL CORPORATION

for the use of the face-hardening process. In large measure because of navy intervention, Harvey was granted a patent for his process, and in the summer of 1891, at the request of the ordnance bureau, he and his superintendent Joseph H. Dickson erected a face-hardening furnace in the armor-finishing No. 3 Machine Shop of the Bethlehem Iron Company. After much experimentation, Bethlehem produced the first commercial face-hardened armor plate in America on July 30, 1892. This Harveyed nickel-steel armor plate was tested at Bethlehem's Redington proving ground. It suc-

cessfully resisted the impact of five 250-pound armor-piercing projectiles fired from an 8-inch gun. All of the projectiles broke apart and penetrated to a depth of only three inches. The combination of nickel steel and face hardening produced what became the basis for almost all of the armor plate produced by both Bethlehem and Carnegie.

Despite the competition of the Carnegie-Phipps armor mill, the Bethlehem Iron Company continued to be a prosperous enterprise. During July of 1892 and March of 1893 Bethlehem received large orders from the navy for

19

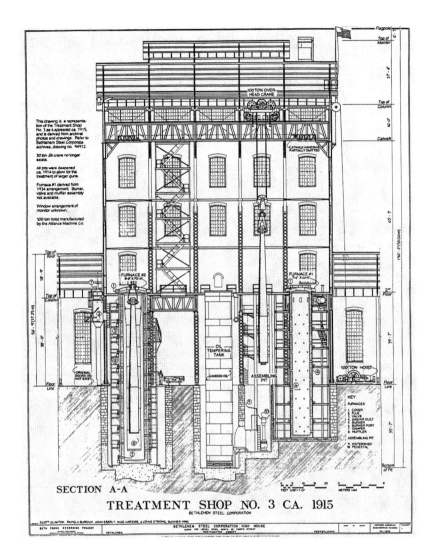

Cross-section of No. 3
Treatment Shop,
c. 1915.

gun forgings and armor plate. To increase their return on the capital investment of the forging plant, the directors of Bethlehem Iron also entered the international ordnance and armor market. In 1894 their efforts were rewarded when the company received a contract for 1,200 tons of armor plate from the Imperial Russian Navy. By 1895 the forging and treatment plant at Bethlehem had become internationally renowned as a leader in the production of ordnance forgings and armor plate. Lieutenant Colonel W. Hope, V.C., a British ordnance expert, visited as a part of his worldwide tour of ordnance works and declared: "I consider the Bethlehem Gun Plant to be superior to any gun plant in the world." The success of American warships during the Spanish-American War

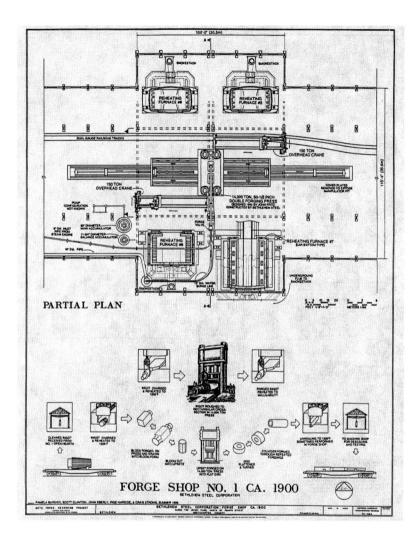

PARTIAL PLAN

FORGE SHOP NO. 1 CA. 1900
BETHLEHEM STEEL CORPORATION

Partial plan of
No. 1 Forge
Shop, c. 1900

served as a great advertisement for Bethlehem Iron since many of the vessels possessed armor, ordnance, propulsion machinery parts, and propeller shafts that had been manufactured by the company. The high regard in which the company was held by the U.S. government became evident when President William McKinley invited Robert Sayre to take a prominent place on the official reviewing stand as the victorious United States Atlantic Squadron passed in a stately procession up the Hudson River on August 20, 1898.

During the remainder of the 1890s Bethlehem Iron continued to broaden the range of its forged products. The company dominated the market for large forgings for the electric power industry and produced the field rings and rotor shafts for the large hydroelectric plants that were being built near Niagara Falls. Its technological expertise

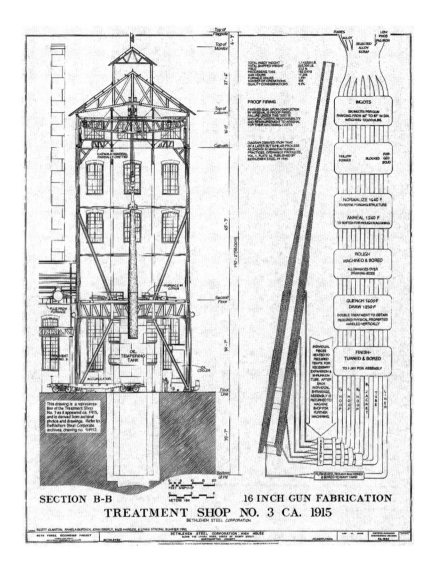

Section of No. 3
Treatment Shop,
c. 1915.

was demonstrated by the forging of important components for the great Ferris wheel of the 1892–1893 Chicago World's Fair; the axle of the Ferris wheel was the largest steel forging to be manufactured up to that date. The company also provided crank shafts and propeller shafts for the engines of many of America's passenger liners and merchant ships.

Design of the Forge Shops

The buildings that John Fritz designed for the forging plant for the Bethlehem Iron Company are unique in that they used John Fritz's innovative shallow-radius, semicircular-section, wrought-iron structural members. Although they resemble Phoenix columns, they differ in many ways from

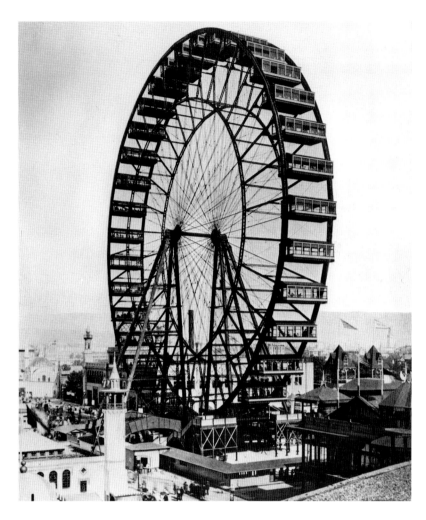

The center axle of the Ferris wheel of the 1893 World's Columbian Exposition in Chicago was forged at the Bethlehem plant.

that widely used structural shape. Phoenix columns, invented and produced by the Phoenix Iron Company in Phoenixville, Pennsylvania, were wrought-iron structural columns that have a modified semicircular central cross-section and flat flanges on the sides. They were commonly used for building bridges. The Fritz-type columns are found only in No. 1 Forge Shop, No. 2 Treatment Shop, No. 1 Open Hearth Shop, No. 3 High House (a vertical quenching and annealing complex), No. 2 Machine Shop,

and No. 3 Machine Shop. No other buildings in the world use this Fritz column as a part of their structure because Fritz retired after the buildings at the Bethlehem plant were completed.

Fritz also deviated from his previous practice of design-ing buildings that were entirely independent of the produc-tion machinery they housed, as in the three-high mill. The buildings that comprised the forging plant had structural frameworks that supported heating furnaces, forging and

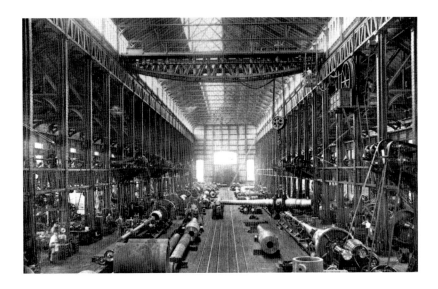

No. 2 Machine Shop, 1,779 feet in length, was the main ordnance facility of the Bethlehem plant.

bending presses, steam hammers, lathes, milling machines, cranes, and hoists. These great buildings had brick exterior walls instead of walls made from local stone. In scale, they dwarfed all other earlier buildings of the Bethlehem plant: No. 2 Machine Shop was over 1,779 feet in length and was considered to be longest and largest machine shop in the world.

The construction of the forging plant between 1885 and 1893 more than doubled the size of the manufacturing complex. All of the great buildings of the forging complex housed active operations until the 1990s.

The Bethlehem Steel Company

The continued success of the Bethlehem Iron Company as a forging manufacturer convinced many of its principal shareholders that they could greatly increase their fortunes by selling all their shares. On April 17, 1899, the directors therefore organized the Bethlehem Steel Company. This was a holding company that immediately leased the properties of the Bethlehem Iron Company for a term of 999 years. It was capitalized at $15,000,000 and immediately issued 300,000 shares of stock, each at a par value of $50.00. The directors of the holding company then offered the stockholders of the Bethlehem Iron Company—essentially the same group of people—the option to buy the shares of the holding company at a ratio of two shares for every one that they held of Bethlehem Iron. The price of the Bethlehem Steel Company's stock was initially pegged at $1.00 per share. This offer allowed the principal shareholders of the Bethlehem Iron Company to gain, for a minimal cash outlay, many additional shares of the Bethlehem Steel Company's stock. The entire transaction was a gift from Bethlehem Iron's capital resources to the stockholders.

The organization of the Bethlehem Steel Company marked the end of an era in the history of iron and steel making at Bethlehem. During the 1890s and early 1900s the principal figures in the company's early development retired or found other employment. This began with the retirement of John Fritz in 1893, continued with the retirement of Robert Sayre in 1898, and concluded with the departure of Russell Davenport in 1902 to assume management control of the Cramp Shipbuilding Company in Philadelphia. However, even before Davenport's departure, ownership of the Bethlehem concern had changed hands.

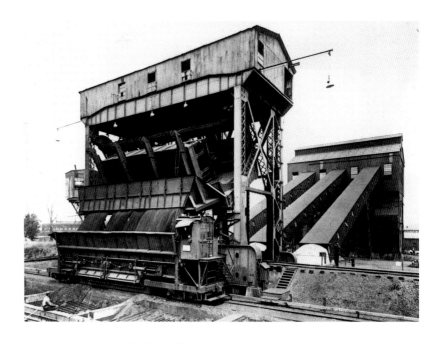

Unloading iron ore into a side-dump car.

In 1901 the British firm of Albert Vickers Sons and Maxim made an offer to buy control of Bethlehem as a means of entering the American ordnance and armor market. Vickers was an immensely powerful conglomerate that produced armor plate, cannons, and warships, and held the principal patents for the self-acting machine gun. On May 28, 1901, Vickers offered to purchase all of Bethlehem's stock at a price of $22.50 per share. The offer was rejected, and a counteroffer of $24.00 per share from Charles M. Schwab, president of the newly organized United States Steel Corporation, was accepted on May 30, 1901. On August 15, the stockholders of Bethlehem Iron voted to accept Schwab's offer. Each of the stockholders received a $1,000 bond for each twenty shares of Iron Company stock that they sold. The lease between the iron company and the steel company was canceled and Bethlehem Iron ceased to exist. In its place was a transformed Bethlehem Steel Company, capitalized at $15,000,000 and with complete operational control of the Bethlehem plant.

Charles Schwab (1862–1939) was among the most brilliant and innovative steelmakers in America. As the protégé of Andrew Carnegie, he had risen to become the president of Carnegie Steel; he personally negotiated the merger of Carnegie Steel with the steel interests of J.P. Morgan to create the United States Steel Corporation, of which he became the first president. Schwab purchased Bethlehem Steel as an independent investment, but soon thought better of it and transferred control to U.S. Steel. However, when approached by a group of investors seeking to create a shipbuilding conglomerate, he was able to repurchase Bethlehem Steel for $7,246,000; he then transferred it to the newly organized United States Shipbuilding Company in return for a large interest in the new concern and placed limits on the amount of control that the shipbuilding company could exercise over Bethlehem.

The United States Shipbuilding Company almost immediately began to experience financial difficulties. Although some of its shipyards were modern, efficient plants, many others were old, obsolete, and had almost no customers

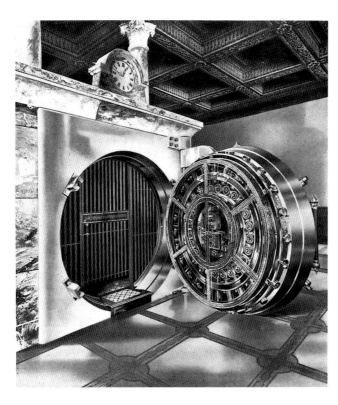

Armor plate bank vault produced at the Bethlehem plant and installed in the Frick Building, in Pittsburgh (Daniel H. Burnham, 1901).

for their limited products. In contrast, the Bethlehem Steel Company continued to be a prosperous enterprise. Its forging plant with its large-capacity machine shop was regarded as the finest in America and was recognized as such in 1902 when it was chosen to build a 12,000-ton forging press and pumping engine for its principal competition, the Homestead, Pennsylvania, plant of the U.S. Steel Corporation. To manufacture these great forging devices, Bethlehem produced the largest and heaviest steel castings yet made, some of which weighed more than one hundred and sixty tons. During this period, the Bethlehem plant employed more than ninety-four hundred workers and maintained a dominant position in the American ordnance and armor market, despite the increased competition brought about by the commencement of armor production by the Midvale Steel Company.

The Bethlehem Steel Corporation

Between 1887 and 1904, Bethlehem produced 42,433 tons of armor plate for the U.S. Navy and an additional 12,500 tons for foreign purchasers. By comparison, Carnegie-Phipps, which was not a major competitor in international sales, produced 46,605 tons of armor for the navy. Bethlehem also produced more ordnance forgings than did the Midvale Steel Company. Partly because Schwab refused to allow a major proportion of Bethlehem's profits to be diverted to the parent corporation, the United States Shipbuilding Company failed in 1903. A series of acrimonious lawsuits from former United States Shipbuilding shareholders followed. Nonetheless, Schwab was able to regain complete ownership of Bethlehem Steel while at the same time salvaging the stronger of his shipyards. He folded these properties into the Bethlehem Steel Corporation,

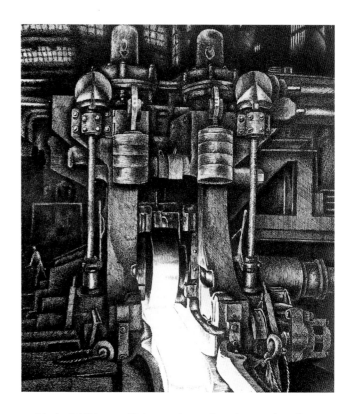

Etching of the forge press from *Fortune* magazine, April 1941.

which was organized on December 10, 1904. Schwab immediately became president of the reorganized company and divested it of several shipyards that were of little value. What remained formed the basis of a greatly enlarged concern that was to become one of the leading U.S. corporations of the twentieth century.

Even before the Bethlehem Steel Corporation was formally organized, Schwab stated his great plans for the company:

> I intend to make Bethlehem the prize steel works of its class, not only in the United States, but in the entire world. In some respects, the Bethlehem Steel Company already holds first place. Its armor plate and ordnance shops are unsurpassed, its forging plant is nowhere excelled and its machine shop is equal to anything of its

kind. Additions will be made to the plant rather than changes in the present process of methods of manufacture. (Hessen, *Steel Titan*)

Schwab believed that a steel company should be managed aggressively to continually seek ways to cut costs and prices, with a resulting increase in market share. This business philosophy clashed with the more conservative attitudes of Judge Elbert Gary, who was chairman of the board of U.S. Steel. In the ensuing power struggle, Schwab was forced out of U.S. Steel in 1903. Since Schwab had helped create U.S. Steel, he was intimately familiar with the company's strengths and weaknesses, and he planned to exploit whenever possible the market opportunities created by U.S. Steel's conservative management policies. Schwab intended to use the Bethlehem plant as

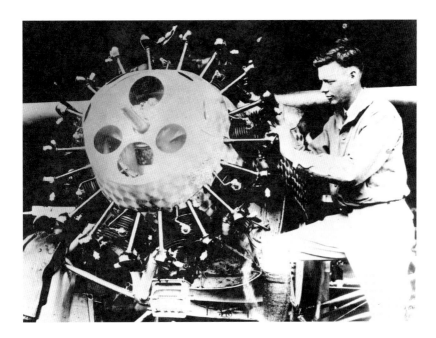

Charles Lindbergh
inspects the Wright J5
Whirlwind engine on the
Sprit of St. Louis; key parts
of the engine were forged
at the Bethlehem plant.

the centerpiece for a large steelmaking concern that could compete successfully with the U.S. Steel Corporation.

Product Diversification

Schwab realized that Bethlehem's dependence on military contracts was, in the long term, dangerous to the company's prosperity. He realized that the new corporation had to develop an expanded line of civilian products to cushion the shock of a sudden downturn in government orders. Among his first acts was to order the installation of a crucible steel plant to produce high-quality alloy steels, and an open-hearth steel rail mill. Because U.S. Steel did not possess an open-hearth rail mill, Schwab could charge a premium for the superior products of this new manufacturing facility and, because his competitor was loath to scrap its large investment in Bessemer steel rail mills, Bethlehem faced little threat in this market from its much larger rival.

Schwab extended the range of Bethlehem's forging activities by adding a large-capacity drop-forging operation during 1905. The drop-forge facility soon won a reputation for the high quality of its products and during the 1920s it pioneered in the production of forged cylinders for the newly developed radial air-cooled aircraft engines. Until the 1980s the drop-forge operations were a major profit center for Bethlehem Steel.

Schwab's belief that Bethlehem Steel would continue to grow was reflected in the architecture of the buildings that he erected to house the new manufacturing facilities. Unlike other steel executives, who constructed cheap but functional buildings from a structural steel framework covered by steel sheeting, Schwab insisted on using brick for trim with decorative concrete or carved limestone for the exterior walls of all new structures built in Bethlehem. He justified this additional expense by remarking to a business associate that "brick buildings are built to last just as I hope that the Bethlehem Steel Corporation will be well managed so that it will survive and prosper." The solidity and appearance of these brick-walled buildings symbolized the corporate image that Schwab wished to project to both the business world and the general public.

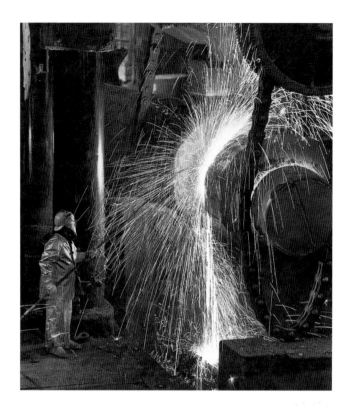

Molten steel is transferred by pouring or "teeming" from the ladle into ingot molds.

The Grey Beam

Schwab's boldest decision was to enter Bethlehem into the growing but fiercely competitive structural steel market. Once again, his knowledge of the strengths and weaknesses of U.S. Steel was a decisive factor in his planning. Schwab knew that the Homestead plant of U.S. Steel was among the largest and most productive manufacturers of structural steel shapes in the nation and that it would be foolhardy for Bethlehem to compete directly with this colossus. As he had done with the installation of an open-hearth rail mill, Schwab planned to produce a superior product to fill a new market niche that was uncontested by U.S. Steel. He found this in 1905 when he committed his corporation to the production of the continuously rolled wide-flange beam.

The continuously rolled wide-flange beam was the development of immigrant British engineer Henry Grey (1846–1913). Grey's innovation was a mill that rolled wide-flange beams directly from ingots. His beams were wider, stronger, and less likely to bend than conventional beams. They were also much cheaper to produce because they could be continuously rolled as a single section, thus eliminating the high costs of riveting and other fabrication that were essential to the production of conventional beams. Grey developed his revolutionary process at the Ironton Structural Steel Company of Duluth, Minnesota, and in 1902 he installed his first full-scale structural mill at Differdingen Steel Works in Luxembourg. Schwab learned of Grey's success while he served as president of U.S. Steel, but Judge Gary did not believe the beam could be mass produced and thus rejected it. Schwab secured the rights to Grey's invention for Bethlehem in 1908.

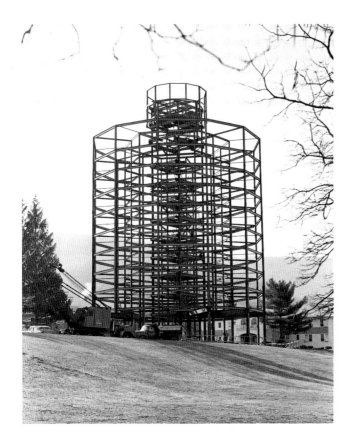

The Nevele Grande resort hotel in Ellenville, New York, under construction in 1965, used Bethlehem structural steel.

Schwab's decision was a potentially dangerous financial gamble for Bethlehem. An investment of almost $5,000,000 was needed to build the new division of the plant that would produce the continuously rolled wide-flange or Grey beam, also sometimes called the Bethlehem or H beam. During the next year Schwab attempted to raise this sum from a variety of sources and by July of 1908 the Saucon Division of the Bethlehem plant with its open-hearth furnaces and structural mill was placed in full operation. Cutting costs and raising production were largely the responsibility of Eugene Grace (1876–1960), who became Schwab's chief protégé and his eventual successor at Bethlehem Steel. However, despite the successful production of the Grey beam, the new product initially found few buyers and Schwab was forced to turn to Bethlehem's forging operations for financial salvation.

Since domestic orders for ordnance and armor plate were flat, Schwab hoped to increase foreign sales of military hardware. Bethlehem was a charter member, along with Krupp, Schneiders, and Vickers-Armstrong, of the international armor-plate cartel named the Harvey United Steel Company, Ltd. This company was a patent pool that held almost all of the important patents concerning armor-plate production; it also served as an informal means by which its members divided up the international armor-plate market. Schwab sought to increase Bethlehem's share of this lucrative trade so that he could gain the funding needed to subsidize the operation of the new

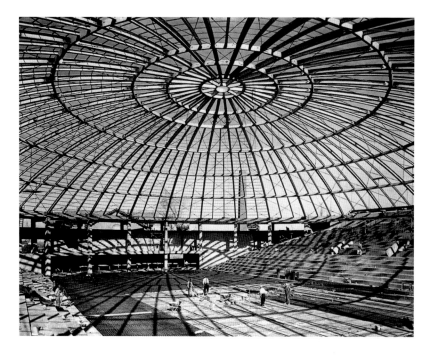

Dome over the ice rink of the George Meehan Auditorium at Brown University, constructed with Bethlehem beams.

Grey mill. Archibald Johnston (1865–1947), Bethlehem's vice president of sales, who had long been associated with the forging operation, was sent to Europe to negotiate with the other members of the Harvey Company. Johnston was successful: by 1908, Bethlehem's share of the international armor-plate market had risen to $2,000,000 annually.

Increased foreign military sales enabled Bethlehem to continue marketing the Grey beam. By 1909 the beam had entered the marketplace, ensuring the success of Schwab's efforts to diversify Bethlehem's product line. By 1914 sales of structural steel were double the annual total value of Bethlehem's forging sales, and construction of the Grey mill doubled the size of the Bethlehem plant.

Although the Grey beam was not an overnight success, it did prove appealing to a number of prominent architects and engineers. These individuals recognized that using the Grey beam would result in substantial savings in construction labor costs since it eliminated the use of overlapping sections of structural steel that were riveted together to form a single beam.

The first large orders for Grey beams were placed by the builders of the New York State Education Building at Albany, and for the construction of a large sugar refinery in Boston. A notable early use of the Grey beam was the Fritz Engineering Laboratory at Lehigh University in South Bethlehem. Completed in 1911–1912, it soon gained an international reputation as a center for the study of structural engineering. The first building to receive widespread notice for its use of Grey beams was Gimbel Brothers' Department Store in New York City.

Architect Ernest R. Graham was an early exponent of the Grey beam and specified its use in a number of notable buildings in Chicago. Among them are the Merchandise Mart, for many years the largest commercial structure in the world, the Marshal Field Office Building, the Field Museum of Natural History, and the Chicago Opera House.

Grey beams provided the framework of such early New

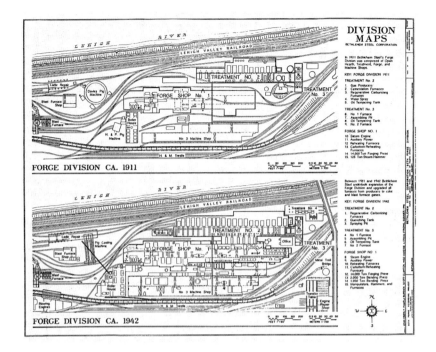

Forge Division
c. 1911 (top)
and c. 1942
(bottom).

York skyscrapers as the Metropolitan Life Insurance Company Tower and the Woolworth Building. Grey beams were also used in the First National Bank Building in St. Paul, Minnesota; the Baltimore Trust Building in Baltimore, Maryland; and the New England Mutual Life Insurance Building and the General Post Office in Boston. One of the great Eastern cities, Philadelphia, boasted no early skyscrapers because city ordinances did not permit any buildings higher than the hat on the statue of William Penn atop City Hall.

The World Wars

The outbreak of World War I in August 1914 was a windfall for Bethlehem Steel. Possessing the largest-capacity forging plant in the United States and already playing a major role as an international supplier of military hardware, Bethlehem Steel was in a unique position to fill orders from the warring powers. The wartime sales bonanza was particu-larly welcome—Bethlehem had not paid dividends since 1906 because of Schwab's consistent policy of reinvesting profits in capital expansion. The influx of war orders also spurred a dramatic rise in the price of Bethlehem's common stock from a level of $30 per share in 1913 to $600 per share in January of 1915, and to an eventual peak of $700 in 1916.

Bethlehem Steel was the first U.S. firm to receive war materials orders from the allied powers of Britain and France; by December of 1914, it had received over $50,000,000 in ordnance orders from these nations in addition to a total order of $135,000,000 for items such as shells and submarines. Unable to fulfill the orders for submarines because of the U.S. position of maritime neutrality, its plants and shipyards nonetheless produced the necessary steel and components and had the submarines assembled in Montréal. To fill the flood of British and French orders, Bethlehem embarked upon a

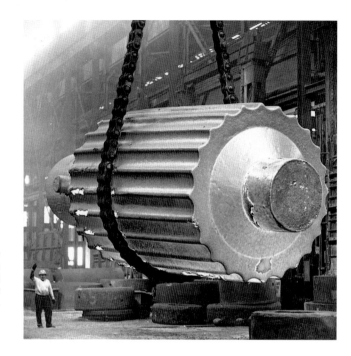

This electric generator shaft—the largest in the world—was produced in No. 2 Forge Shop for General Electric in the 1950s.

rapid expansion of its production facilities. In 1913 the total work force of the Bethlehem plant was 9,000 people. By the end of 1914 this number had grown to 24,567 employees, of whom more than 2,000 were employed on plant construction alone. Over $25,000,000 was spent on expansion, including the completion of a large, four-story addition to No. 2 Machine Shop, the company's primary heavy-ordnance finishing facility. By the end of 1914 the products of Bethlehem's forging operation were being used by many other parts of the plant to produce military products, such as No. 3 Machine Shop, which specialized in the manufacture of field artillery caissons, and No. 4 Machine Shop, which became Bethlehem's primary producer of field guns. Other shops combined forces to turn out more than 2,000 shells per hour; these shells were filled with high-power explosives at the new loading facilities located at the company's Redington proving grounds almost two miles to the east of the main plant.

Despite a disastrous fire that destroyed No. 6 Machine Shop in 1915, temporarily decreasing military production, the output of war materials from the Bethlehem plant became ever greater during 1915 and 1916. When the United States entered World War I in 1917, the prominence of Bethlehem Steel as a military contractor rose to new heights. By 1918 over 35,000 workers were employed at the plant, including large numbers of women for the first time in company history. By 1919 the plant had produced 60 percent of the finished guns ordered by the United States, 65 percent of all American gun forgings ordered, and 40 percent of the nation's artillery shell orders. At the same time, the plant was producing steel for the company's shipyards, the largest in the world, which were working overtime to provide the troop carriers and merchant ships needed to carry personnel and goods to France. In addition, forges and machine shops supplied the French armed forces with semi-finished gun tubes for more than 21,000 field pieces. For Britain and France combined it

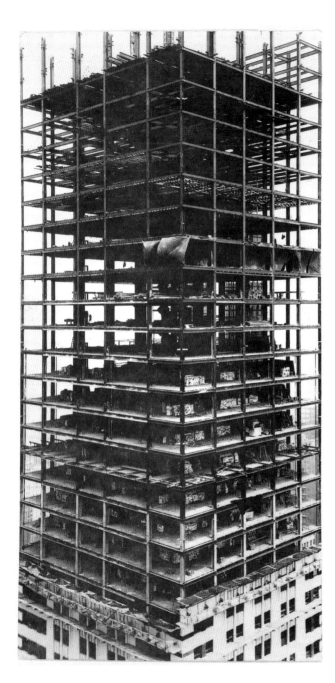

The H beams used in the construction of the Chrysler Building in 1929 were produced at the Bethlehem plant.

supplied 65,000,000 pounds of forged military products, 70,000,000 pounds of armor plate and an incredible total of 1,100,000,000 pounds of steel for shells, and 20,100,000 rounds of artillery ammunition. Between April of 1917 and the Armistice in November of 1918, Bethlehem produced more than 65 percent of the total number of finished artillery pieces manufactured by all of the allied nations. To achieve these miracles of productivity, the company had expended more than $102,000,000 for the construction of new facilities at Bethlehem.

The enormous profits earned by Bethlehem Steel during World War I enabled Schwab to undertake an acquisition program that included many of the remaining independent American steelmakers. The first, and in many ways most notable, of these were the Pennsylvania Steel Company and its subsidiary, the Maryland Steel Company, acquired on February 16, 1916. During 1916 Bethlehem also purchased the American Iron and Steel Company of Lebanon, Pennsylvania, a firm that held a large share of the American nut and bolt market, and the large Cornwall, Pennsylvania, iron mines of the Lackawanna Iron and Steel Company. In 1919, Bethlehem purchased control of the Midvale Steel Company in order to acquire its subsidiary, the Cambria Steel Company at Johnstown, Pennsylvania. Fearing antitrust action, Bethlehem did not purchase Midvale's Philadelphia plant, which became the focal point of a reorganized and independent Midvale concern. Bethlehem's final major purchase occurred on October 9, 1922, when it gained control of the Lackawanna Steel Company with its massive plant located at Lackawanna near Buffalo, New York.

This massive acquisition program made Bethlehem a well-rounded competitor to the U.S. Steel Corporation in most areas of commercial steel production. Between 1905 and 1925 its steelmaking capacity expanded from an ingot capacity of 190,000 tons annually to 7,000,000 tons annually. Equally remarkable was the eventual decline in the amount of Bethlehem's military-related production after World War I. In 1905 approximately 92 percent of the cor-poration's annual production was devoted to military pro-duction; by 1925 less than 5 percent of its products con-sisted of such items. The increasing popularity of the Grey beam and other commercial products absorbed almost all of the facility's productive capacity by then.

Between 1919 and 1921, the Bethlehem Steel Corpora-tion completed what may be considered to be the most architecturally distinguished of the Bethlehem plant's buildings: the powerhouse. This building housed a series of 25-cycle AC generators and blowing engines powered by furnace gas. These mechanical marvels, based on a design that the Mesta Corporation, a Pittsburgh steel machinery manufacturer, had imported from Germany, produced electrical power for the entire plant and provided the "wind" or pressure that in turn provided the blast for the blast furnaces. Jacobean-style motifs executed in carved limestone and cast cement sculptural panels decorated this magnificent steel-frame and brick structure.

A Turning Point Between the Wars
The Washington Naval Conference of 1922 had a devastat-ing impact on the Bethlehem Steel Corporation. The result

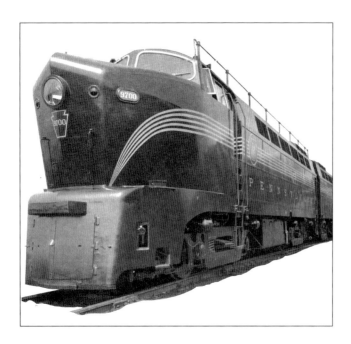

Pennsylvania Railroad
locomotive, c. 1950s,
made with Bethlehem-
patented Mayari Steel.

was the virtual cessation of battleship construction in the United States and a drastic reduction in the building of other armored warships. The signing of the Washington Treaty brought about an abrupt halt to production on February 9, 1922. All phases of operations on all unfinished armor plates were immediately halted. The plates that were closest to completion were shipped to navy storage depots and all others were scrapped. Bethlehem received a large settlement from the navy for the losses that it incurred from the canceling of these contracts. No further armor contracts were received until December 1930. When the manufacture of armor plate resumed in 1931 after almost nine years of virtual inactivity, nearly all of Bethlehem's armor-plate production facilities were in poor condition. Production capacity, then only 500 tons per month, soon expanded due in large measure to the gradual buildup of the navy in response to what was perceived as a growing threat from the Japanese fleet.

The U.S. Navy completed design work on a new class of battleships (the USS *North Carolina* and the USS *Washing-*

ton) in 1937, giving renewed momentum to a steadily accelerating program of new ship construction. With this increase in naval construction and the corresponding rise in government orders for ordnance, armor plate, and propulsion machinery parts, Bethlehem Steel began to expand and upgrade its forging and treatment operations. Improvements continued until the tremendous production demands of World War II required a much greater expansion of Bethlehem's forging and treatment facilities.

During World War II, Bethlehem Steel built many important defense plants and additions to existing structures involved in war production. Among the largest of these projects were the mammoth Chrysler Tank Arsenal near Detroit, Michigan, and the three-quarter-mile-long Consolidated Vultee bomber plant at Fort Worth, Texas. Both of these massive plants remain in operation. Chrysler Tank, owned by the U.S. government, is today managed by General Dynamics Land Systems and still produces tanks for the U.S. Army, one of only two plants in the nation with this capability. Consolidated Vultee is currently owned by

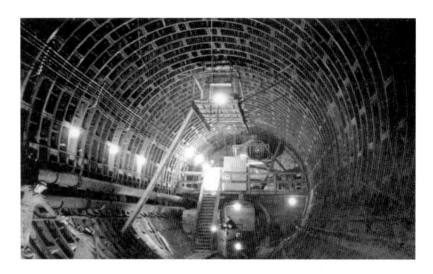

The lining of the Brooklyn Battery Tunnel, opened in 1950, includes iron segments cast and machined at Bethlehem.

Lockheed Aerospace and produces F-16 fighter planes for the United States and allies.

New wartime buildings at the actual plant, such as the No. 2 Forge Shop, brought to an end the long-established tradition of building for permanency using brick for exterior walls. By the late 1930s, Bethlehem Steel was using steel frames and corrugated steel sidings for its buildings. These the new forge shops and other buildings had a much more utilitarian construction of a simple steel framework covered by steel sheeting. The most extensive of the new facilities was the East Lehigh complex. Henceforth, no more brick-walled building would be built at the Bethlehem plant.

The Bethlehem Steel Corporation was one of the largest defense contractors in the United States during World War II. Its yards had built over 1,121 merchant and naval vessels and repaired an additional 38,000 ships, and in 1944 it had made more than 13,000,000 tons of raw steel. This massive production meant that the U.S. armed forces had many vessels that were suddenly surplus to its needs after the defeat of Japan in 1945. For more than a decade production of almost all major warships ceased; in the case of battleships this meant the permanent cancellation of all vessels of this type that

were under construction. The halt to construction of heavily armored warships virtually shut down Bethlehem's armor plant in 1946 and, with the exception of a single experimental plate produced for the navy in 1956, no forged armor was made at this facility.

The Postwar Years
Building America: Fabricated Steel Construction
The period between 1945 and 1973 was one of unprecedented prosperity. Since its 1931 purchase of the McClintic-Marshall Corporation, Bethlehem had been one of the largest fabricators and erectors of skyscrapers and long-span bridges in the world.

Among the notable buildings constructed by the Fabricated Steel Construction Division (or FSC) of the Bethlehem Steel Corporation in the New York metropolitan area prior to 1948 are the Chrysler Building, the Stuyvesant Town Apartment Complex, the Waldorf-Astoria Hotel, the New York Central Building, Century Apartments, Rockefeller Center (including the Associated Press Building), the Tishman Realty Company Building, and the Tiffany Building. During the great postwar building boom, Bethlehem Steel continued to make many notable additions to the New York City skyline through the construction of the Coli-

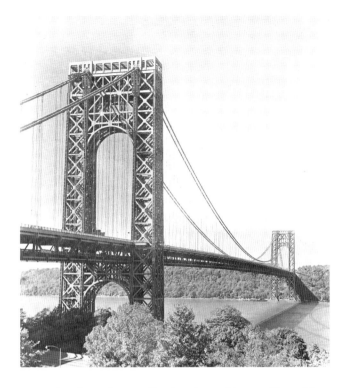

FCS was responsible
for the construction
of the George
Washington Bridge
in New York City.

seum, Madison Square Garden, Lever House, Chase Manhattan Bank, and the Citicorp Center.

FSC was responsible for many of the important American spans built during the first half of the twentieth century. The most famous of these are the Golden Gate Bridge across the mouth of San Francisco Bay; the George Washington Bridge, which crosses the Hudson River between New Jersey and Manhattan; and the spectacular bridge rising over New York Harbor at the Verrazano Narrows, linking Staten Island with Brooklyn. A shorter but equally famous suspension bridge is the Ambassador Bridge between Detroit, Michigan, and Windsor, Ontario. A Bethlehem-built bridge that is well known to tourists is the Rainbow Bridge across the Niagara River near the Niagara Falls. Several major Bethlehem spans cross the Mississippi River at cities such as Baton Rouge, Louisiana; St. Louis, Missouri; and

Dubuque and Lansing, Iowa. FSC also built such projects as the second suspension span of the Delaware Memorial Bridge near New Castle, Delaware, the Chesapeake Bay Bridge connecting the eastern and western shores of Maryland, and the Commodore Barry and Betsy Ross bridges across the Delaware River near Philadelphia. FSC's last major bridge was the I-95 span across the Piscataqua River between New Hampshire and Maine.

Many highway bridges in the New York metropolitan area were early Bethlehem projects. Among them are the Pulaski Skyway between Newark and Jersey City, New Jersey, the Thomas Edison Bridge across the Raritan River at Perth Amboy, New Jersey, the lift bridge that carries the Pennsylvania Railroad over the Passaic River at Newark, New Jersey, the Mill Basin Bascule on the Shore Parkway at Brooklyn, and the Bronx Kills approach span to the Triborough Bridge.

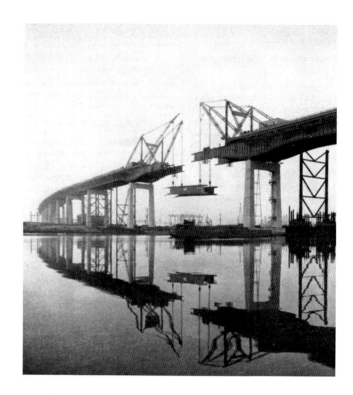

Similtaneous erection of two 146-ton closing girders on the Pasaic River Bridge of the New Jersey Turnpike.

Several major government offices, such as the Department of Labor and Interstate Commerce and the Supreme Court building in Washington, D.C., have steel structural frameworks composed of Grey beams from the Bethlehem plant. In Los Angeles, both the Hall of Justice and City Hall were products of Bethlehem's plants. A great irony is that two of the most famous landmarks in Pittsburgh, home of Bethlehem's greatest rival, U.S. Steel—were constructed from Bethlehem beams. These skyscrapers are the Koopers Building and the world-renowned Cathedral of Learning at the University of Pittsburgh.

Bethlehem also built many significant canal- and navigation-improvement structures such as the mitre gates of the lock adjacent to Bonneville Dam on the Columbia River, the tainter gates of the lock at Dam No. 4 on the Mississippi River near Alma, Wisconsin, and the world-famous lock gates of the Panama Canal.

The 1950s and 1960s: Plant Expansion and Modernization
The end of armor-plate production made the great 14,000-ton-force hydraulic forging press and its associated steam pumping engine largely redundant. In 1952 it was scrapped and the No. 1 Forge Shop was converted to a relining facility for hot-metal transfer railroad cars. During the 1950s Bethlehem developed a new line of forging products as the result of the construction of nuclear-powered warships and power plants. The company was also able to gain a major competitive advantage through the development of the vacuum-degassing method of casting steel forging ingots, in which molten steel is poured from a furnace ladle into a smaller or pony ladle; from there it is drained into an ingot mold contained in a vacuum chamber. As the molten steel passes into the chamber it breaks up into tiny droplets, allowing most of the hydrogen and oxygen gases it contains to escape. By the time it reaches

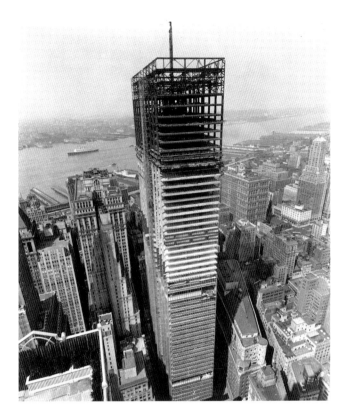

The Chase Manhattan Bank in New York City, under construction in 1959, used 53,000 tons of Bethlehem steel.

the mold the steel is almost free of gas bubbles and as it cools it assumes a largely stress-free internal structure. The vacuum-degassing process, which Bethlehem had used since 1956, and its existing heavy forging capacity, enabled the company to dominate the U.S. market for heavy forging.

During the 1950s the plant again proved its importance to the U.S. armed forces by providing much of the ordnance for the Korean War. In 1957, Bethlehem Steel employed over 165,000 people, its highest level of employment since World War II. The 1950s was a decade of increased prosperity for workers, particularly after a prolonged strike in 1959 resulted in greatly increased levels of wages and benefits. Many workers moved to the suburbs, away from the South Side of Bethlehem and the plant itself. This trend greatly

accelerated a breakdown that affected almost all of the ethnic communities that had once clustered together in the shadow of the mill. Social halls still stand in Bethlehem, but their members now come in ever decreasing numbers, from all over the Lehigh Valley.

Recognizing the importance of both Midwestern markets and the growing dominance of sheet and plate products, the management of the Bethlehem Steel Corporation developed plans for further expansion. In 1962, Bethlehem Steel announced that it would build a major integrated steel plant at Burns Harbor, Indiana. This new plant was the single most ambitious undertaking in the corporation's history. In 1964 the new plant was placed in operation with an annual production capacity of more than 5,300,000 tons of steel sheet and plate. Burns Harbor was

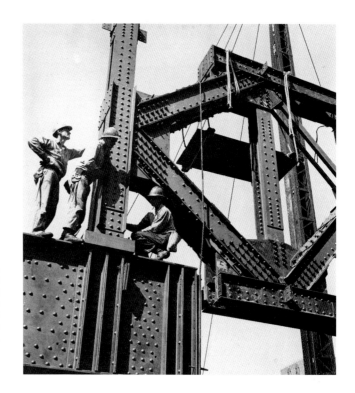

Ironworkers tie in a 53-ton, 70-foot-long truss on the Sheraton Hotel in Philadelphia. The structural steel framework was fabricated and erected by Bethlehem Steel.

the last fully integrated steel plant to be built in the United States and, by the 1970s, it became the Bethlehem Steel Corporation's major profit center.

The decision to construct the Burns Harbor plant was a primary example of the symbiotic relationship that was developing between the automobile industry and American steel companies during the 1960s and 1970s. In many ways this relationship mirrored the formative role that railroads had played in the creation of the steel industry a century earlier. Burns Harbor was designed to produce sheet steel, the primary component of automobile and appliance bodies, and its location was chosen to be near the factories of Detroit and the other automobile manufacturing centers of the Midwest.

The 1960s was an era of major capital expenditures and enlargement at the Bethlehem plant. In 1967, the corporation replaced the open-hearth steelmaking furnaces with a European innovation, the basic oxygen furnace. The basic oxygen furnace shop was built on a 36-acre site in what was known as the Northampton Heights redevelopment area. It was housed in a steel-frame building with a corrugated sheet steel exterior, 905 feet long in the east-west direction with maximum width of 308 feet. The highest point of the building towered over 256 feet over the surrounding landscape. The principal components were two 250-ton-capacity basic oxygen furnaces, which were mounted on trunnions and could be rotated around a horizontal axis to facilitate charging and pouring.

The basic oxygen steelmaking system was originally developed in Austria. In many respects it resembled the earlier Bessemer process, but instead of air being blown into the furnace vessel, pure oxygen was blown into the vessel at supersonic speed using a water-cooled lance. The refractory-lined steel furnace vessel was charged with a

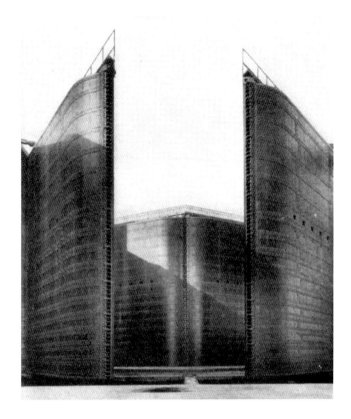

The McClintic-Marshall Corporation, later a division of Bethlehem Steel, fabricated the Panama Canal lock gates; each of the 12 leaves is 110 feet high.

mixture of molten pig iron and scrap before the blowing occurred. The basic oxygen process possesses several advantages over the open-hearth process and, in relative terms, they are cheaper to build and maintain than open-hearth furnaces. Basic oxygen furnaces greatly reduced the amount of time needed to produce steel, making them much more efficient than the widespread open-hearth furnaces. Upon its completion in 1967, the basic oxygen steelmaking plant became the primary steelmaking facility at the Bethlehem plant.

The 1970s: Fabricated Steel Construction Closes
The high-water mark for the Bethlehem Steel Corporation and the end of an era for the Bethlehem plant came in 1973. During that year Bethlehem Steel produced 22.3 million tons of raw steel and shipped 16.3 million tons of finished steel. Both of these figures were company records. In 1973 Bethlehem Steel made a net profit of $207,000,000. But such good times were not fated to last.

During the 1970s, competition from cheaper imported steel and declining export markets brought about a sharp decline in both production and earnings. In 1976 Bethlehem's management shut down its Fabricated Steel Construction Division. Rising costs and competition from smaller contractors using cheaper imported steel forced management's decision. The closure of FSC had a major negative impact on the Bethlehem plant. Because FSC was one of the principal customers for structural steel produced at Bethlehem, its closing caused a domino effect that brought about the shutdown of other operations.

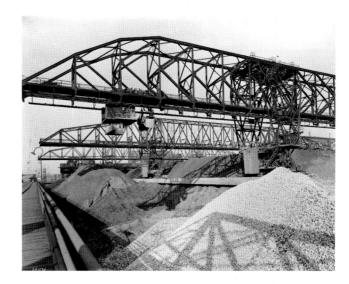

Ore-loading bridge cranes with iron ore stockpiles in the foreground.

Losses resulted in reduced steelmaking capacity at both the Johnstown and Lackawanna plants in 1977. This decision set in motion a process that would eventually result in closure or sale of the production facilities at those plants. In 1982 Bethlehem Steel reported a record loss of more than $1.5 billion, the first of five years of losses as management implemented a massive restructuring plan involving the shutdown and sale of plants, mills, mines, and shipyards. The corporation's total employment was reduced by 50 percent. Increasingly, the Bethlehem Steel Corporation became dependent on its operations at Burns Harbor, Indiana, and Sparrows Point, Maryland, as its hopes for future profitability.

The 1980s and 1990s:
Restructuring and the End of the Bethlehem Plant
Mini-mills presented formidable competition during the 1980s and 1990s. These mills, which produce steel by remelting scrap steel in electric arc furnaces, were able to produce steel at far lower costs than integrated steelmakers such as Bethlehem. The most aggressive of the new competitors was the Nucor Corporation, which in 1988 opened a structural mill in Arkansas. Nucor could acquire

scrap, make new steel products, ship it to the Lehigh Valley, sell below the price that Bethlehem Steel charged, and still make a profit.

By 1990 the outlook for the Bethlehem plant became increasingly grim: 6,000 people were employed in 1990; by 1995 this figure fell to 4,000. Plans were announced to drastically downsize the plant and install a large-capacity electric arc furnace and essentially convert the plant into a mini-mill. But the economics of this plan could not be made to work, and so it was decided to end steelmaking at the Bethlehem plant.

On Saturday, November 18, 1995, steelmaking officially ended in Bethlehem. Although plans had been made to continue to operate the combination mill with steel produced at the Steelton plant and shipped to Bethlehem, this operation did not succeed and within a year the combination mill was also shut down.

The last part of the Bethlehem plant to operate was the coke works. It had been hoped that the coke works would be able to remain profitable as an independent operation. This plan did not succeed and on March 28, 1998, the coke works was shut down, ending all of the Bethlehem Steel Corporation's manufacturing activities in Bethlehem.

43

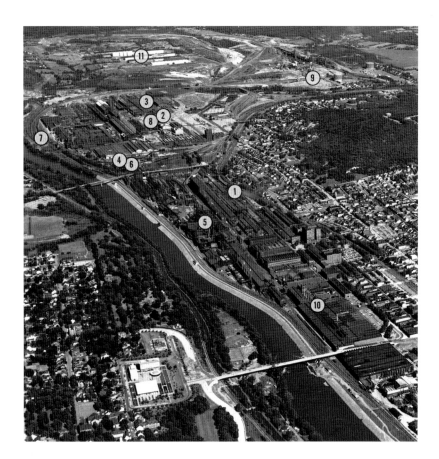

Aerial view of the plant in 1962; the plant covered four and a half miles along the Lehigh River. The numbers indicate various pollution abatement facilities.

Much of the now-idle Bethlehem plant is being scrapped, but plans have been announced to save a significant part of it as a multifaceted museum. The Bethlehem Steel Corporation has created a nonprofit subsidiary, the Bethlehem Works, to plan and manage the reuse of the site. Among the most interesting aspects is the conversion of the monumental No. 2 Machine Shop into a National Museum of American Industry as a joint venture of the Bethlehem Works and the Smithsonian Institution. Other portions of the plant such as the blast furnaces and powerhouse will become part of a Steel Heritage Museum, and such historically important buildings as the original Bessemer steel plant will be adaptively reused for other purposes. In this way future generations will have the opportunity to understand the magnitude of the role the steel industry played in our national development, in a location of outstanding historic significance.

Bethlehem Steel Timeline

1857   The Sauconna Iron Company is formed but never takes any action.

1860   Lehigh Valley Railroad executive Robert H. Sayre takes control and reorganizes the company as the Bethlehem Rolling Mill and Iron Company. The reorganized company will produce iron rails. Sayre hires John Fritz, inventor of the three-high rail mill, to design and manage the plant.

1863   Bethlehem Rolling Mill and Iron Company changes its name to the Bethlehem Iron Company. Production of rails begins at the plant located in South Bethlehem, along the Lehigh River at the junction of the North Pennsylvania and the Lehigh Valley railroads.

1873   Production of Bessemer steel rails begins at the Bethlehem plant.

1887–1894   Construction of massive plant for the manufacture of heavy forgings such as ordnance, armor plate and propulsion machinery parts. The Bethlehem Iron Company becomes the birthplace of the modern American defense industry and military-industrial complex. John Fritz retires.

1898   American warships with Bethlehem components win the key battles of Manila Bay and Santiago de Cuba, enabling the United States to be victorious in the Spanish-American War and become a world power.

1899–1901   The Bethlehem Steel Company is organized as a holding company for the Bethlehem Iron Company. Bethlehem Steel enters the international ordnance and armor market. Robert Sayre retires.

1901–1904   Charles M. Schwab, president of the United States Steel Company, purchases control of Bethlehem Steel and merges it with United States Shipbuilding Company. United States Shipbuilding fails.

1904   Out of the Bethlehem Steel Company and the remnants of the United States Shipbuilding Company, Schwab organizes the Bethlehem Steel Corporation.

1907–1908   Schwab builds Henry Grey's revolutionary continuously rolled wide-flange beam mill at the Bethlehem plant. These H beams are lighter, stronger, and more economical than the conventional riveted I beams.

1910   A major and violent workers' strike takes place at the Bethlehem plant. It results in complete management victory.

1914–1918   The Bethlehem plant becomes the single most important source of war materials for the Allied armed forces during World War I. The British liner *Lusitania* is sunk carrying Bethlehem munitions.

1916–1922   Fueled by huge war profits, Bethlehem Steel expands through the purchase of many of the remaining independent American steel makers, bringing under its control plants and shipyards at Johnstown, Pennsylvania; Quincy, Massachusetts; Steelton, Pennsylvania; Lebanon, Pennsylvania; Sparrows Point, Maryland; and Lackawanna, New York.

1920–1930s   The Grey beam, also known as the Bethlehem or H beam, produces much of the framework for America's skyscrapers and long-span bridges.

1931   Bethlehem purchases McClintic-Marshall Corporation, a major steel fabricating and erecting company, and creates the Fabricated Steel Construction Division.

1941   After a short and violent strike, the U.S. government pressures the management of the Bethlehem Steel Corporation to accept worker representation by the United Steel Workers of America.

1939–1945   Bethlehem plant once again becomes a major source of war materials for the Allied effort in World War II. It supplies not only guns, armor plate, propulsion machinery parts, and ammunition, but also the majority of parts needed for the production of aircraft engine parts. Bethlehem Steel Corporation builds more than 1,000 ships for the U.S. and British Navies.

1959   Lengthy strike idles Bethlehem plant and begins cycle of concessions to the union that lead to ever-rising wages and costs.

1964   Bethlehem Steel completes a new greenfield plant on Lake Michigan at Burns Harbor, Indiana, to produce steel sheet and plate. This plant was the last integrated steel mill to be built in America.

1971   Rising energy costs and competition from foreign steel imports and mini-mills that produce steel from remelted scrap cause the Bethlehem Steel Corporation to begin a campaign of downsizing that will eventually lead to plant closings.

1976   The Fabricated Steel Construction Division is closed down, creating a domino effect in other production departments.

1982–1983   The massive Lackawanna plant near Buffalo, New York, is largely closed.

1992   The Cambria plant at Johnstown, Pennsylvania, is closed.

1995–1998   The Bethlehem plant is shut down. Plans are announced to preserve many of its most historic structures as part of the National Museum of American Industry, a showcase of nineteenth- and early twentieth-century production machinery from the collections of the Smithsonian Institution. Other significant structures such as the blast furnaces and power house are designated to be preserved as a separate Steel Heritage Museum.

Bibliography

The major manuscript collections on the Bethlehem Steel Corporation are located at the Hagley Museum in Wilmington, Delaware, and the Archives of the National Canal Museum in Easton, Pennsylvania. The personal papers of Bethlehem's founders, Robert H. Sayre and John Fritz, are located at the National Canal Museum archives.

Books and Papers

Archer, Robert F. *The History of the Lehigh Valley Railroad.* Berkeley, CA: Howell North, 1978.

Bartholomew, Craig L., and Lance E. Metz. *The Anthracite Iron Industry of the Lehigh Valley.* Easton, PA: Center for Canal History and Technology, 1988.

The Bethlehem Steel Corporation. *Half a Century of Fabricated Steel Construction.* Bethlehem, PA: Bethlehem Steel Corporation, 1948.

Billinger, R.D. "Beginnings of the Bethlehem Steel Corporation." *Bulletin of the Commonwealth of Pennsylvania Department of Internal Affairs.* Vol. 20, Feb. 1953.

Brown, Mark M. "The Architecture of Steel: Site Planning and Building Type in the Nineteenth-Century American Bessemer Steel Industry." Ph.D. diss., University of Pittsburgh, 1995.

Coleman, Lyman. *Guidebook of the Lehigh Valley Railroad.* Philadelphia, PA: J. B. Lippincott, 1872.

Cooling, Benjamin Franklin. *Gray Steel and Blue Water Navy: The Formative Years of America's Military Industrial Complex 1887–1917.* Hamden, CT: Archon Books, 1979.

Cotter, Arundel. *The Story of Bethlehem Steel.* New York: Moody Magazine and Book Co., 1916.

Drinkhouse, W. Bruce. *The Bethlehem Steel Corporation: A History from Origin to World War I.* Easton, PA: The Northampton County Historical and Genealogical Society, 1964.

Fritz, John. *The Autobiography of John Fritz.* New York, NY: John Wiley & Sons, 1912.

Hessen, Robert W. *Steel Titan: The Life of Charles Schwab.* New York: Oxford University Press, 1975.

Jaques, W.H. "Description of the Works of the Bethlehem Steel Company." *Proceedings of the United States Navy Institute.* Vol. XV (4) 1889.

McHugh, Jeanne. *Alexander Holley and the Makers of Steel.* Baltimore, MD: Johns Hopkins University Press, 1980.

Metz, Lance E. *Robert H. Sayre, Engineer, Entrepreneur and Humanist.* Easton, PA: Center for Canal History and Technology, 1985.

———. *John Fritz: His Role in the Development of the American Iron and Steel Industry and His Legacy to the Bethlehem Community.* Easton, PA: Center for Canal History and Technology, 1987.

———. "The Arsenal of America: A History of Forging Operations of Bethlehem Steel." *Canal History and Technology Proceedings.* Vol. XI. Easton, PA: Canal History and Technology Press, 1992.

Misa, Thomas J. *A Nation of Steel.* Baltimore, MD: Johns Hopkins University Press, 1995.

Sayenga, Donald. "Canals, Converters and Cheap Steel." *Canal History and Technology Proceedings.* Vol. VIII. Easton, PA: Center for Canal History and Technology, 1989.

Warren, Kenneth. *The American Steel Industry: A Geographical Interpretation.* Oxford: Clarendon Press, 1973.

Yates, W. Ross. *Bethlehem of Pennsylvania: The Golden Years.* Bethlehem, PA: Bethlehem Book Committee, 1976.

———. *Joseph Wharton: Quaker Entrepreneur.* Bethlehem, PA: Lehigh University Press, 1987.

Periodicals

"Bethlehem Steel." *Fortune* XXIII (4): 61–70 (April 1941).
"Steel Capacity." *Fortune* XXIII (4): 70–73 (April 1941).

PHOTOGRAPHS
1995–1999
ANDREW GARN

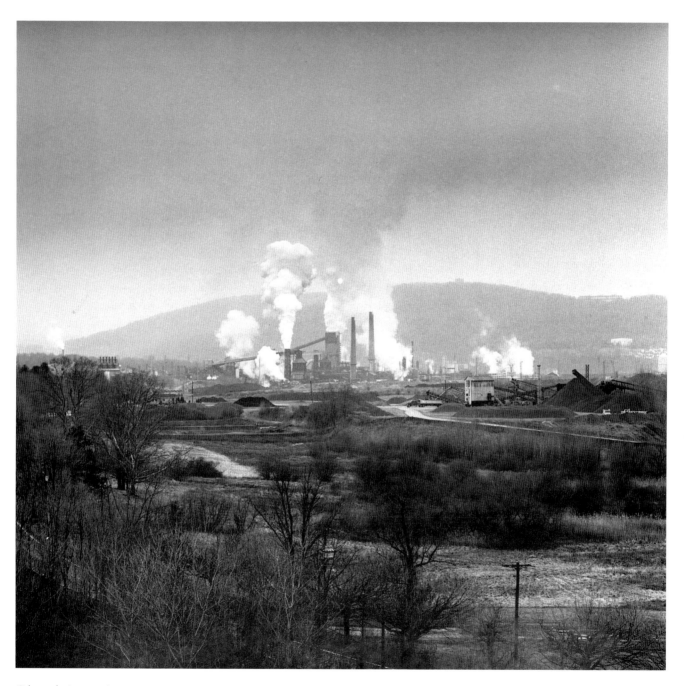

Coke works in operation.

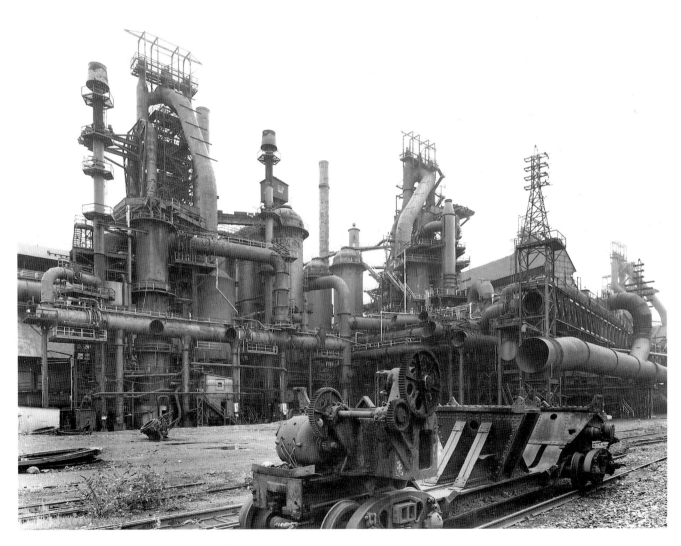

Blast furnaces A, B, C, D, and E with slag cars in foreground.

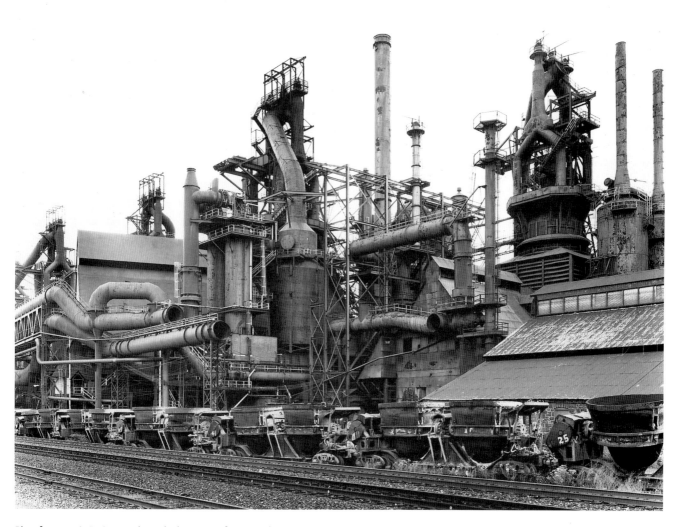

Blast furnaces A, B, C, D, and E with slag cars in foreground.

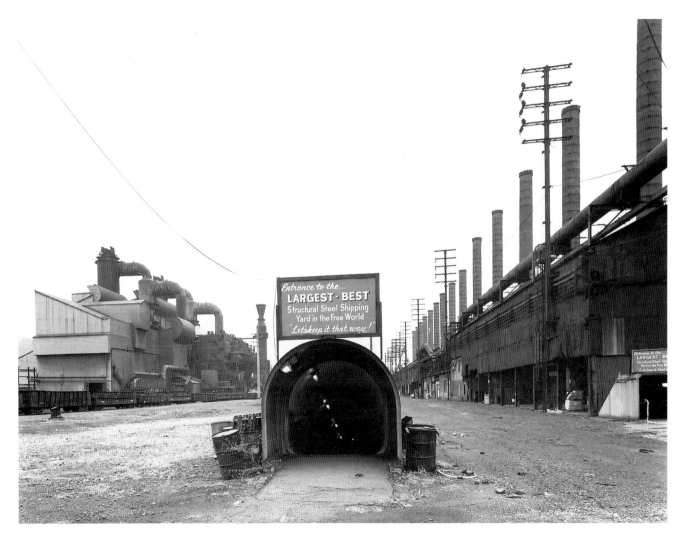

Underground pedestrian entrance for workers.

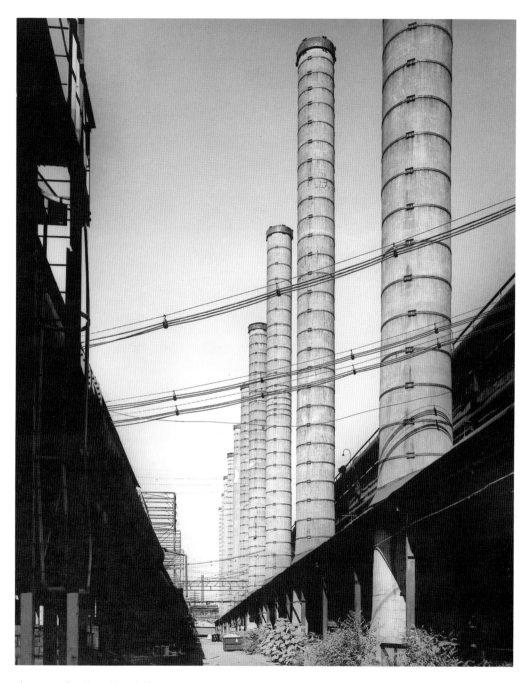

Chimneys of an Open Hearth Shop.

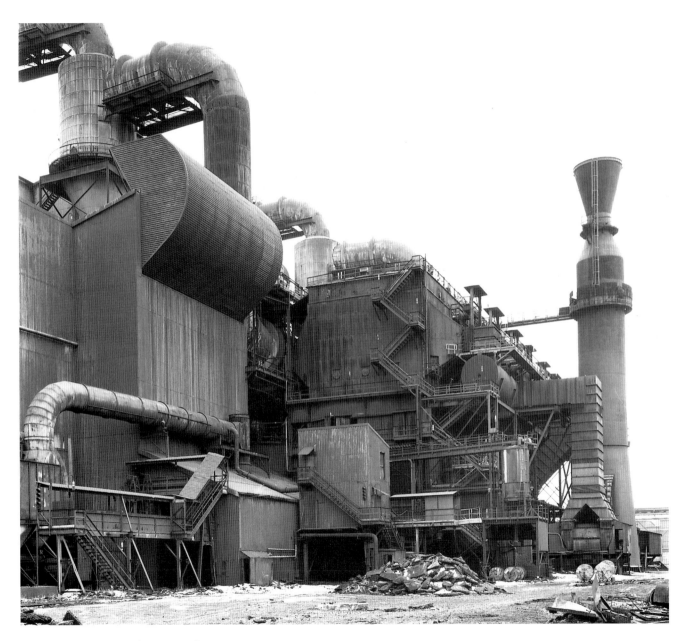

Electrostatic precipitator at basic oxygen furnace.

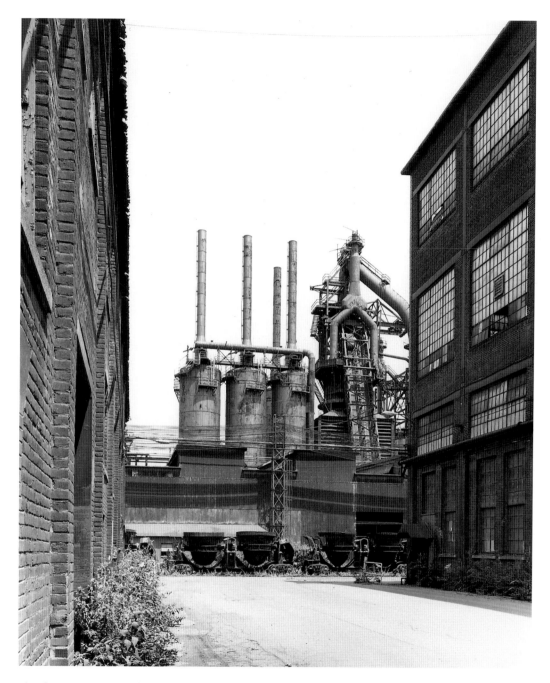

Blast furnace A, near central tool shops.

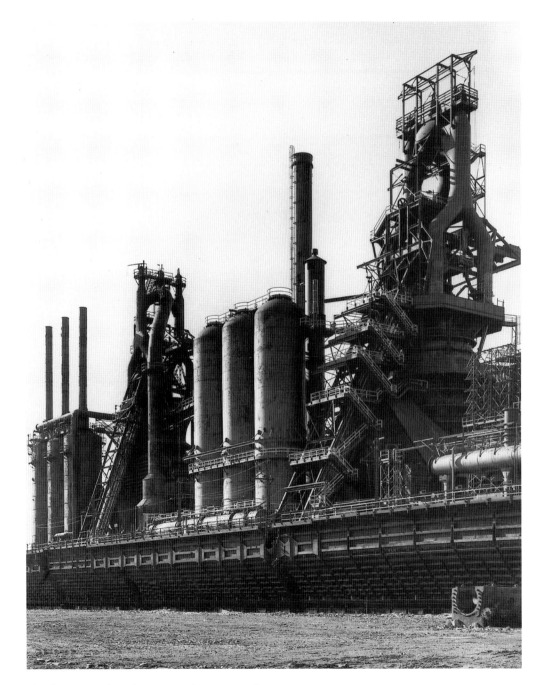

Blast furnace A and B, adjacent power house removed.

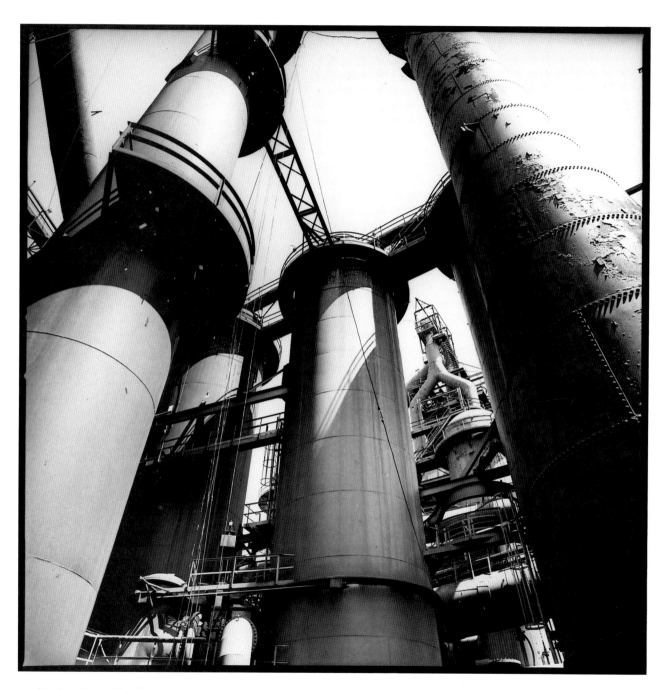

Gas bleeder valve and blast furnaces.

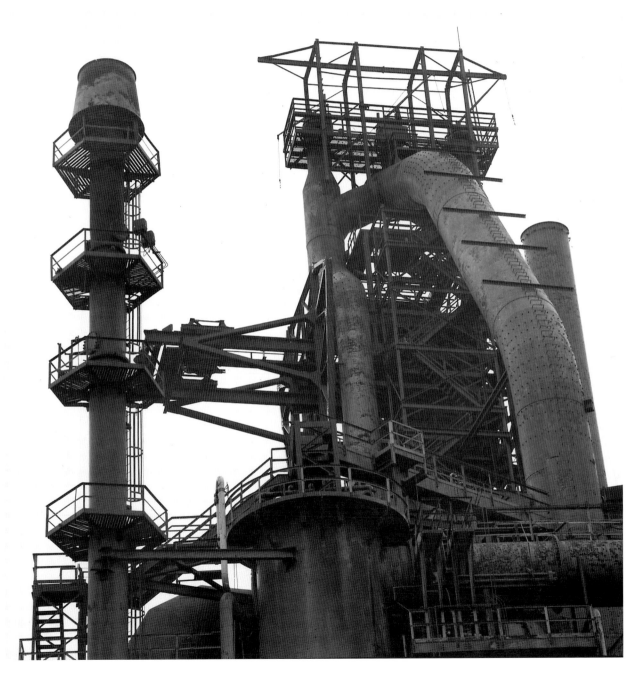

Blast furnace stacks.

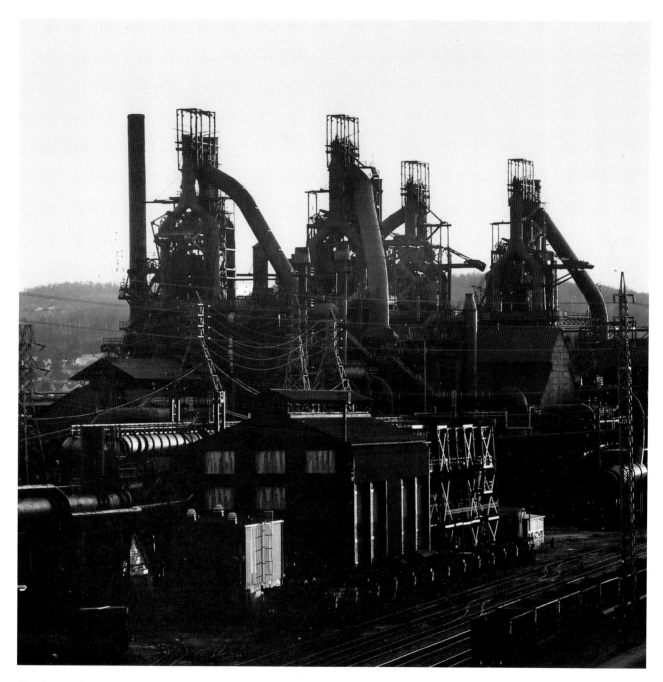

Blast furnace C.

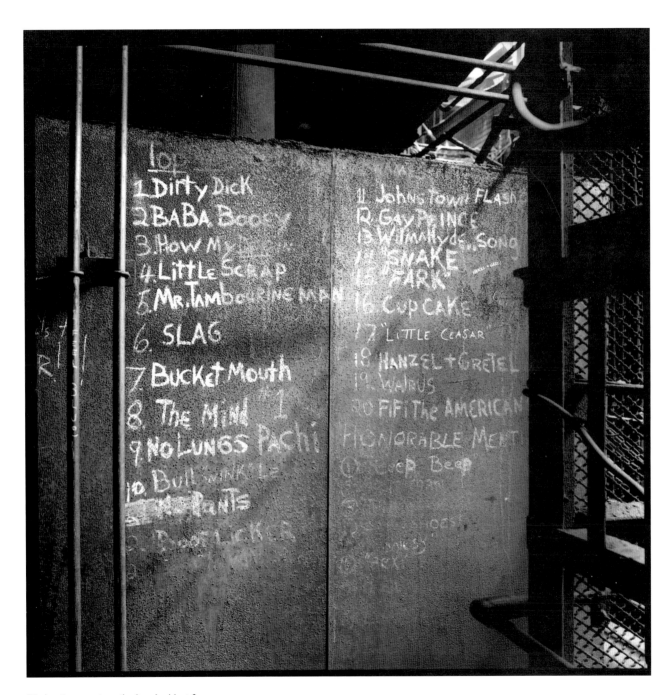

Workers' names inscribed under blast furnaces.

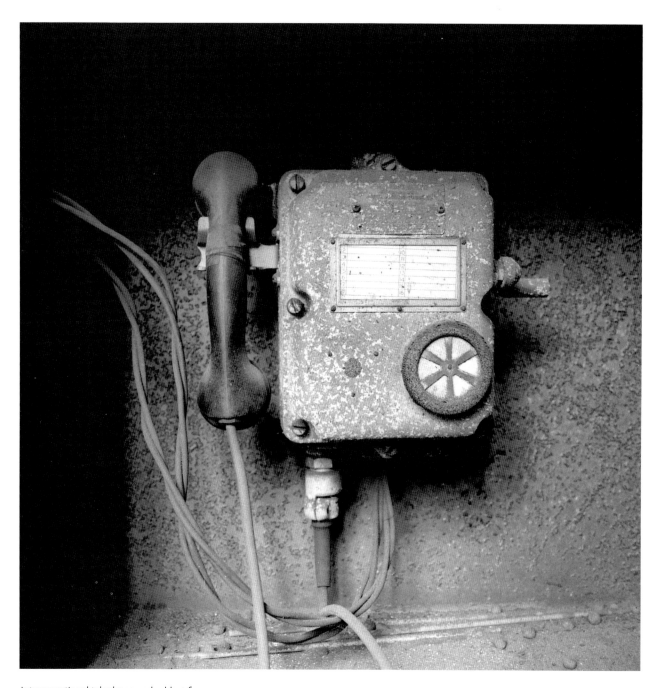

Interoperational telephone under blast furnaces.

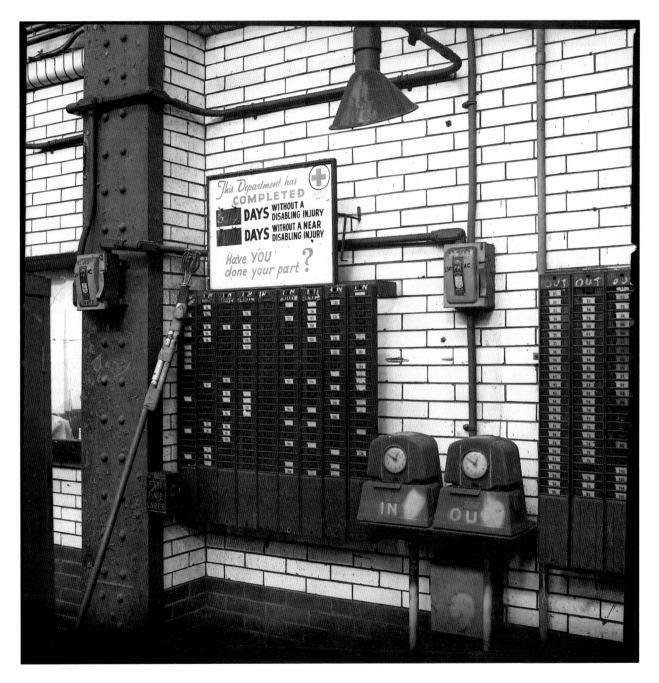

Time card racks in the power house.

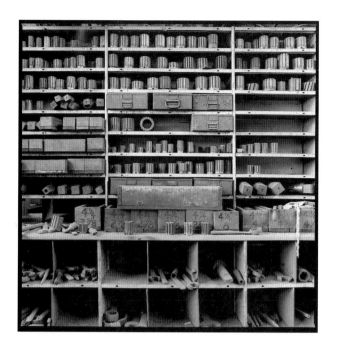
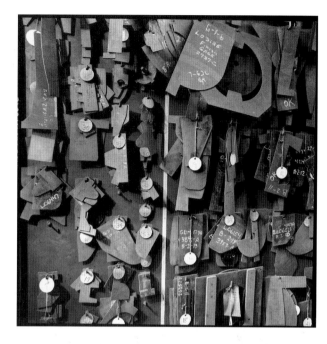
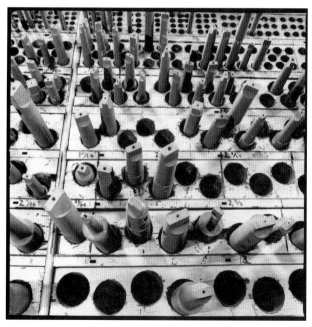
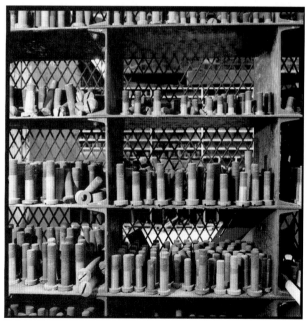

Clockwise from top left: Tool storage in No. 2 Machine Shop;
mechanics templates; bolts in the mechanics' tool room; tool bits in the tool room.

65

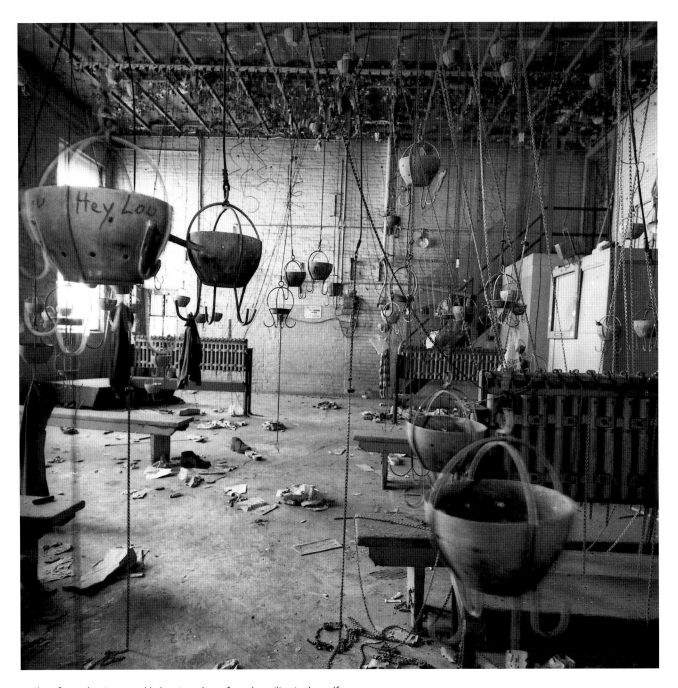

Baskets for workers' personal belongings, hung from the ceiling in the welfare room.

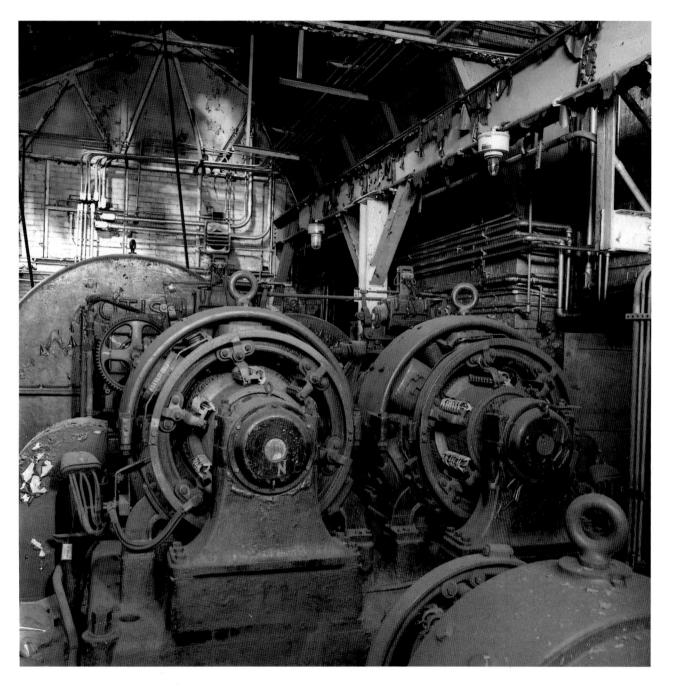

Otis electric motors under the blast furnaces.

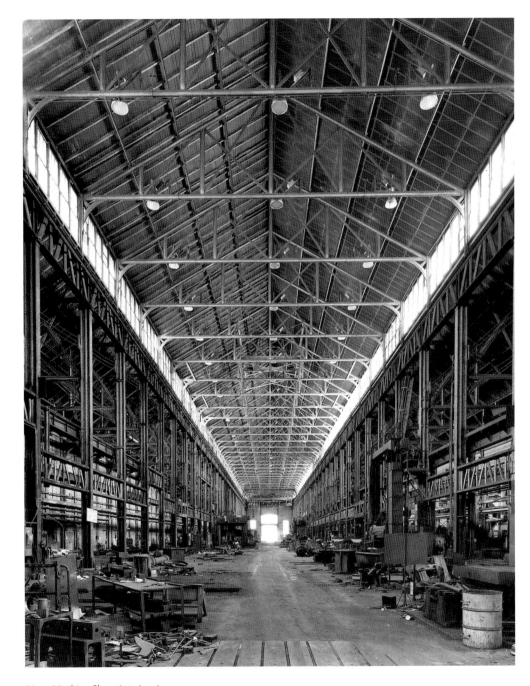

No. 2 Machine Shop, interior view.

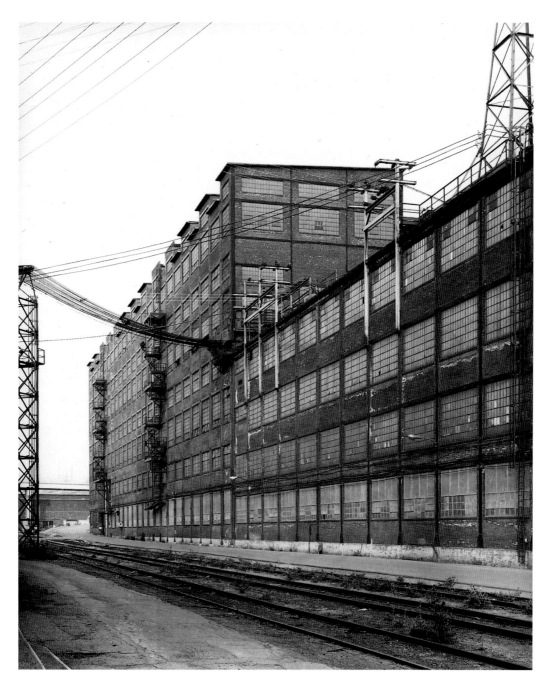

Exterior of No. 2 Machine Shop.

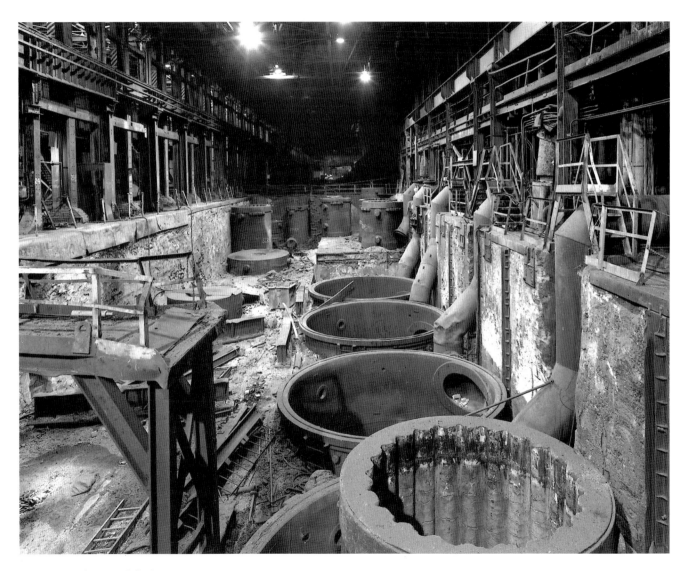

Casting pit at electric melt facility.

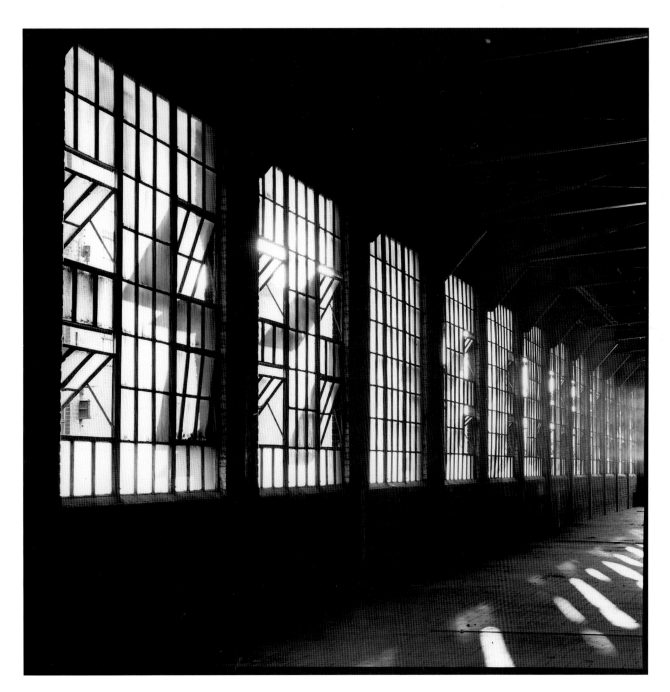

Interior of carpenters' shop.

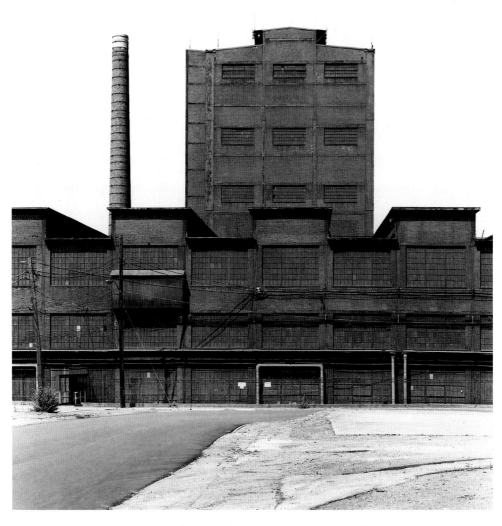

No. 5 High House (vertical Treatment Shop), with the No. 1 Machine Shop in the foreground.

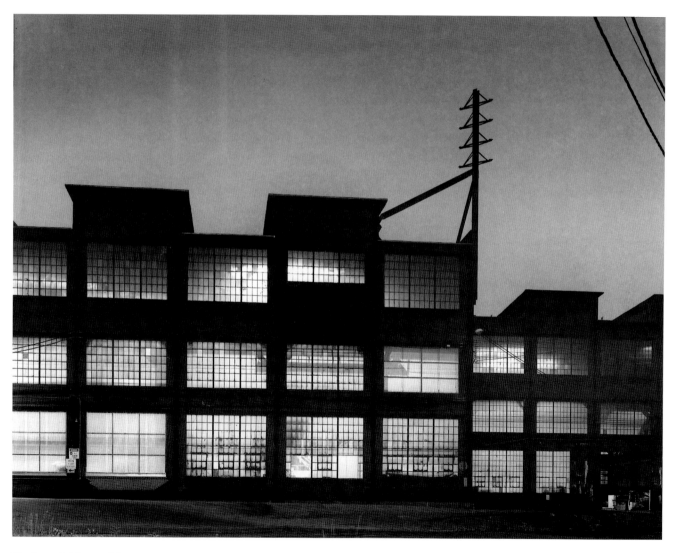

The forge division at night.

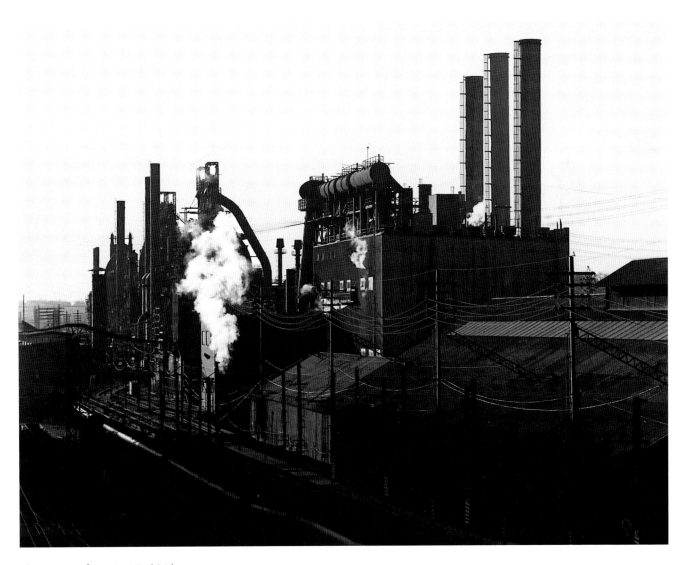

Plant overview from Minsi Trail Bridge.

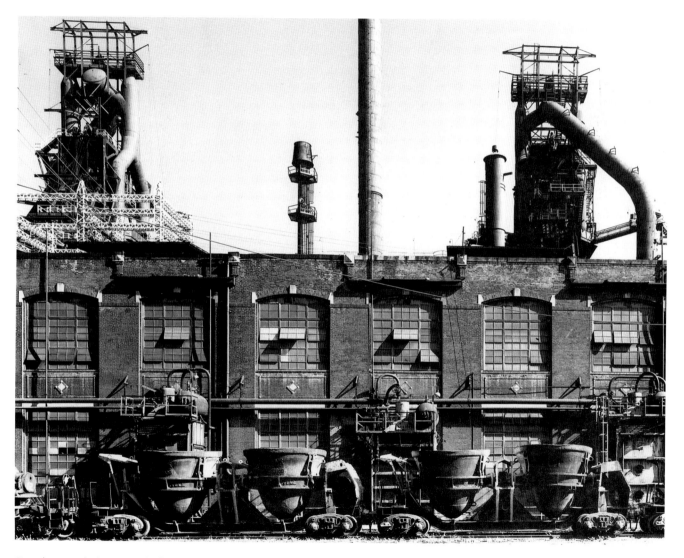

Powerhouse with slag cars in the foreground.

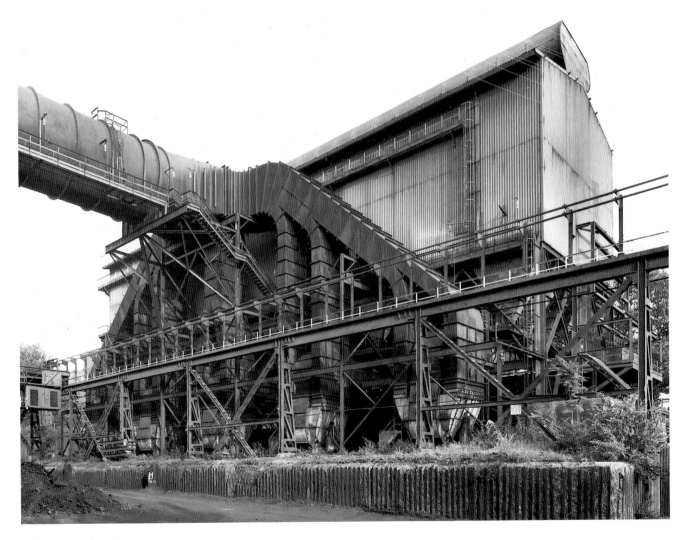

Pollution-control equipment.

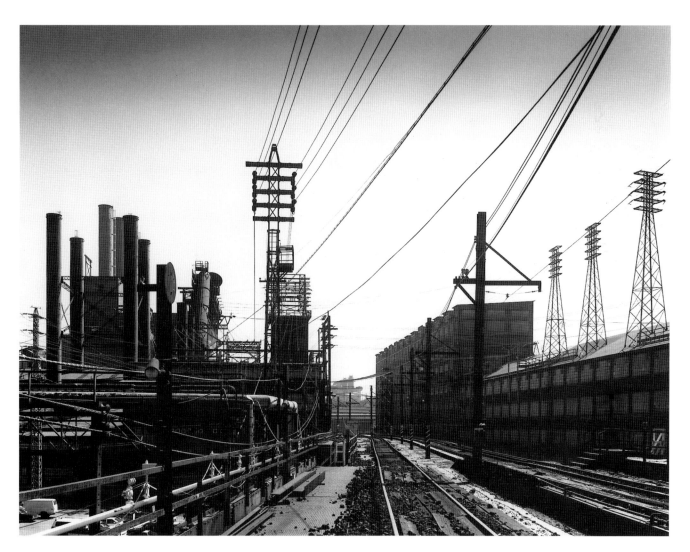

Railroad trestle looking east.

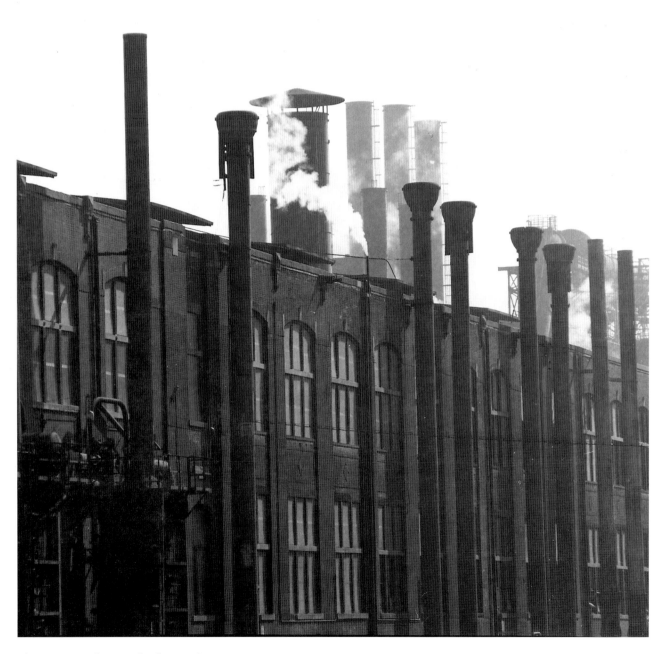

Blowing engine exhaust stacks of a power house.

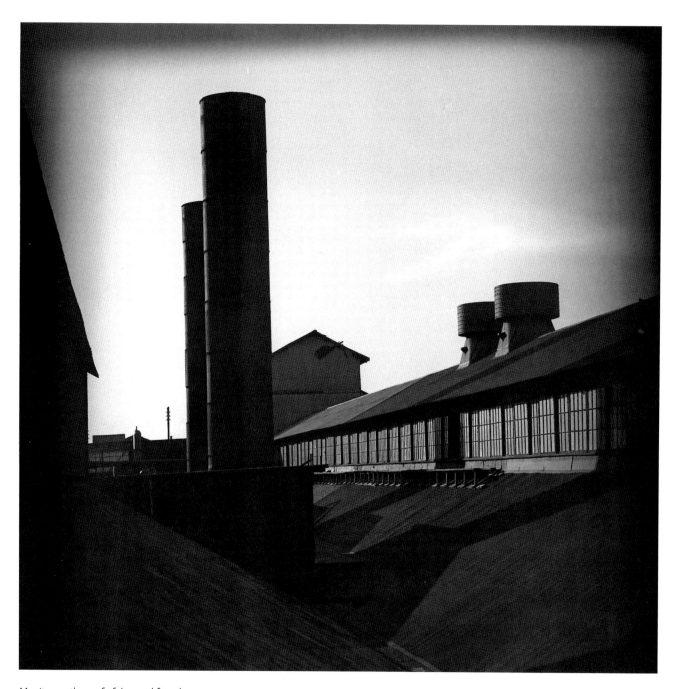

Monitor on the roof of the steel foundry.

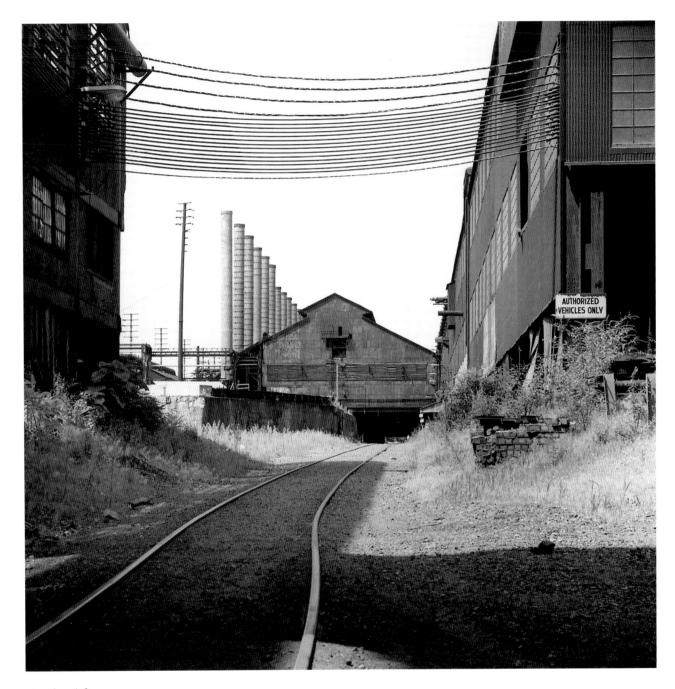

Open-hearth furnaces.

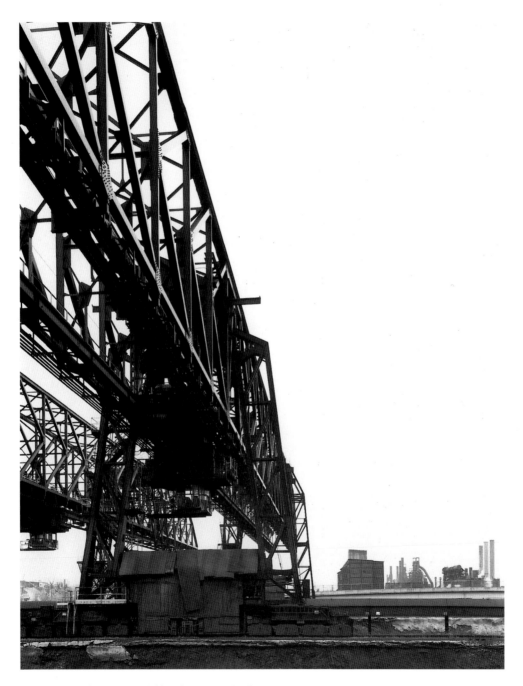

Ore-loading bridge cranes with blast furnaces in the distance.

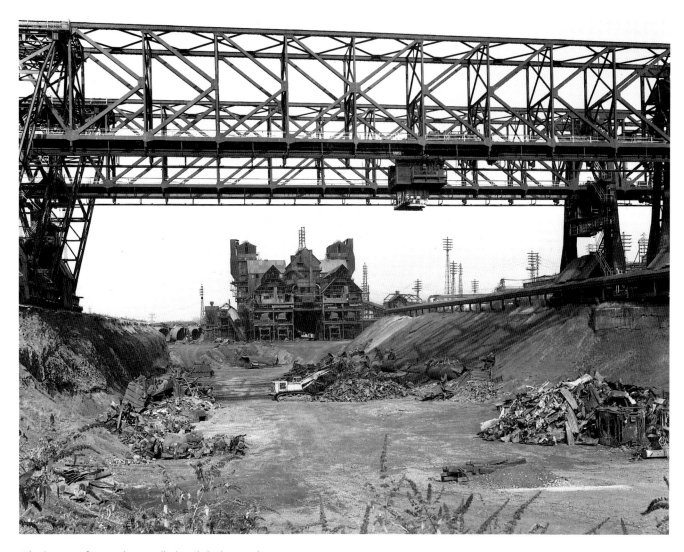

A bridge crane framing the partially demolished sinter plant.

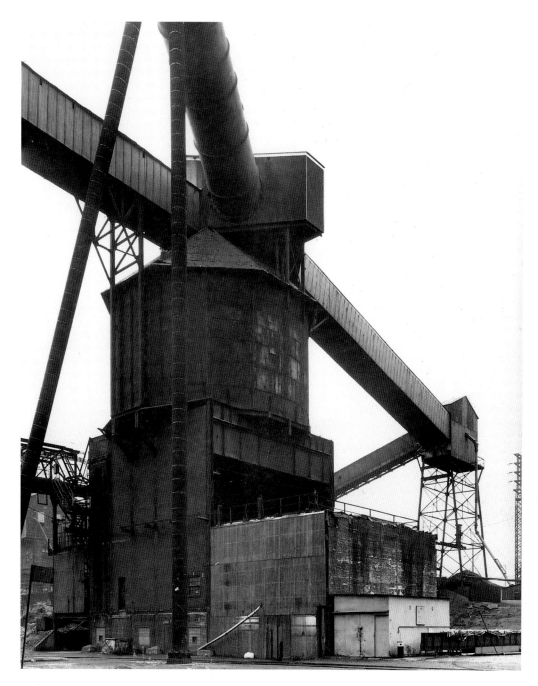

Coal conveyors at the coke works.

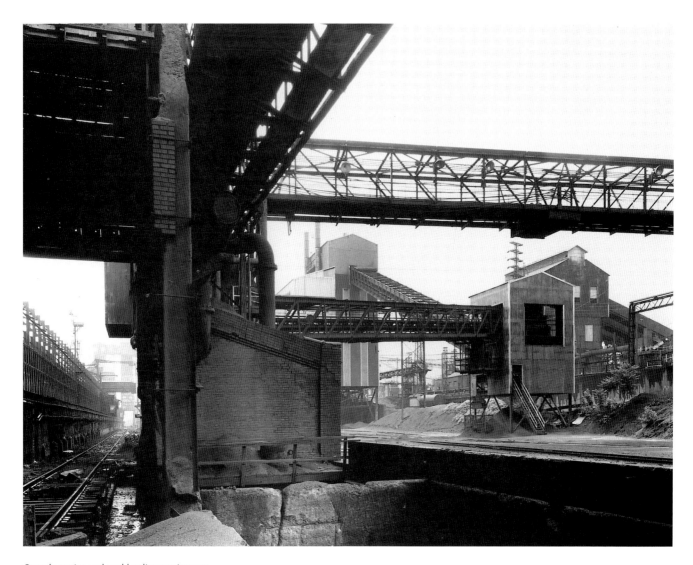

Oven batteries and coal-loading equipment.

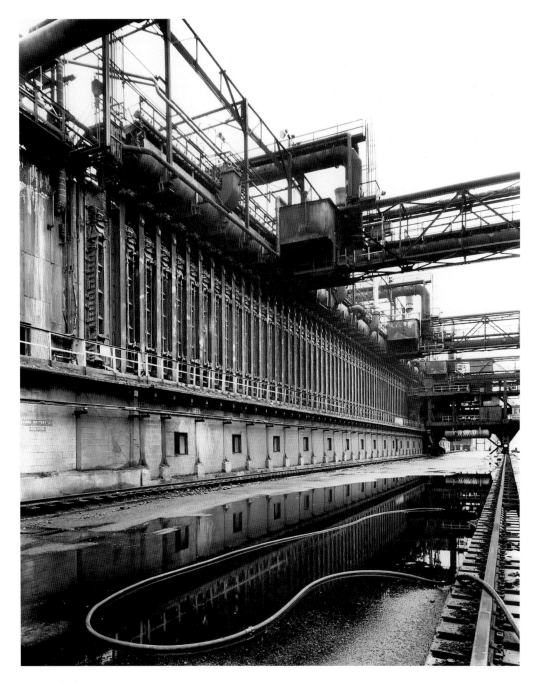

Battery of coke ovens.

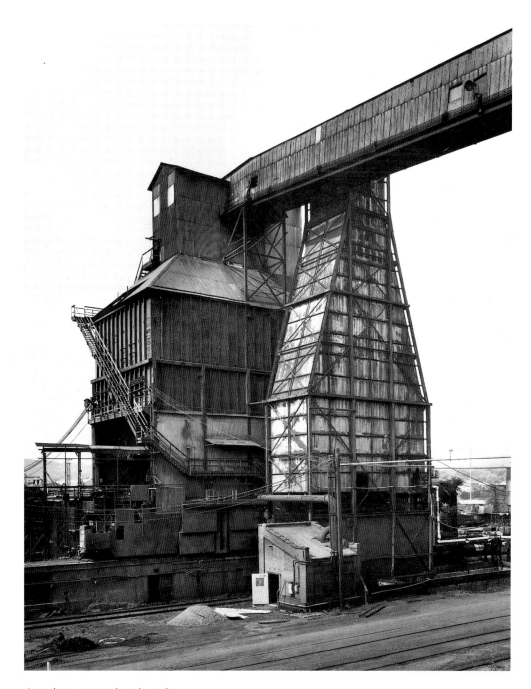

Quenching tower at the coke works.

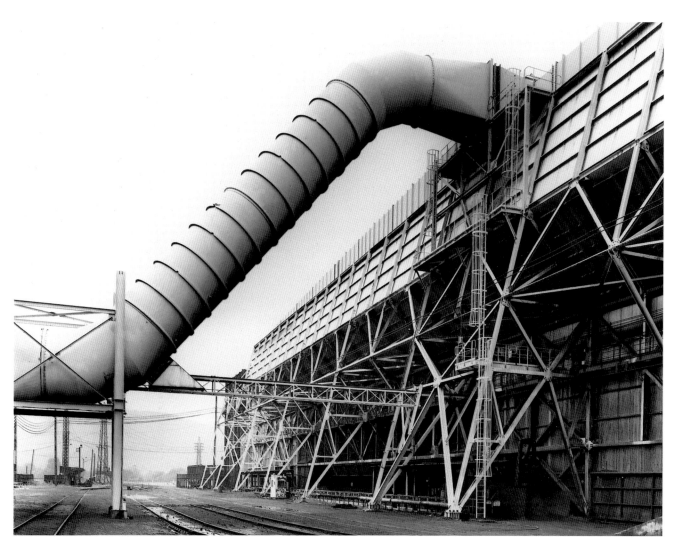

Emission-containment system covering the coke oven battery.

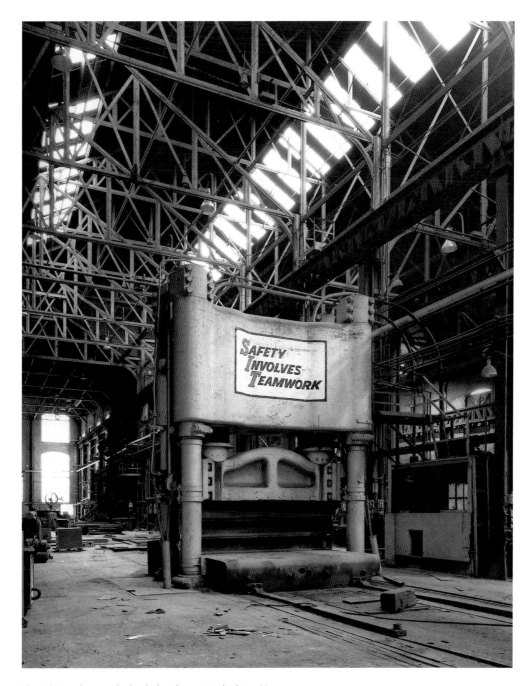

The Whitworth steam hydraulic bending press, built in 1887.

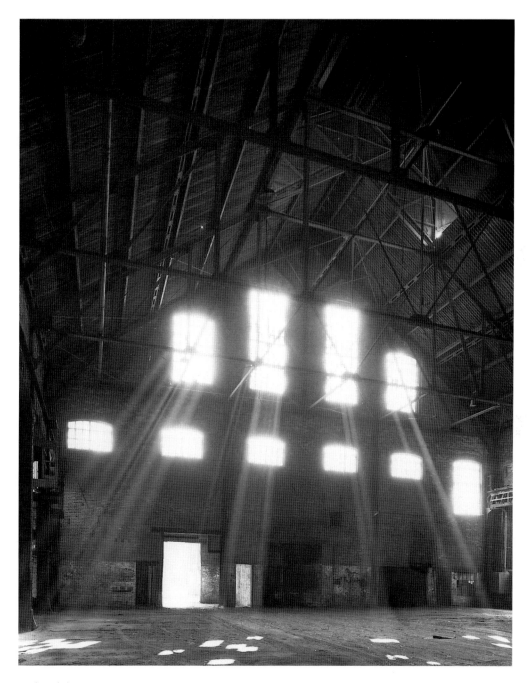

Tool steel shop interior.

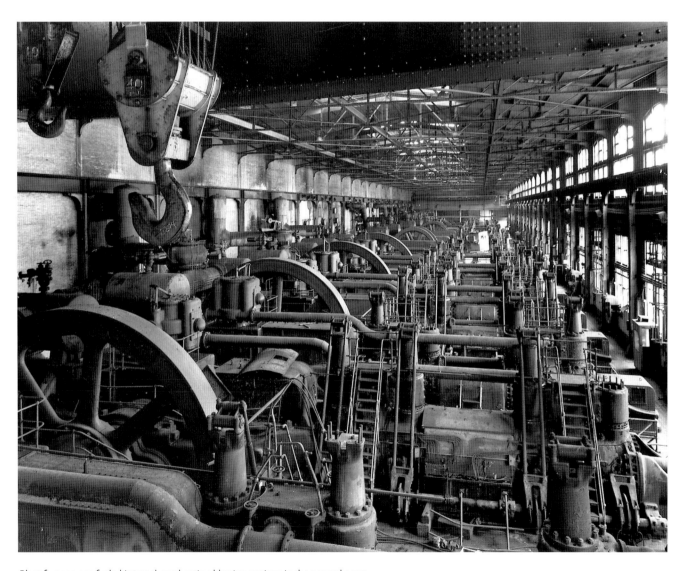

Blast-furnace-gas-fueled internal combustion blowing engines in the power house.

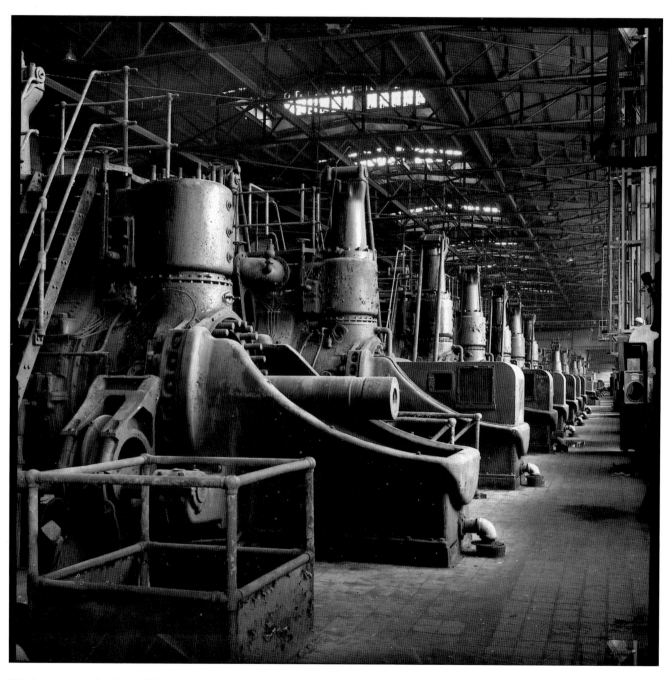

Tail piece and cross-head end of the blowing engines.

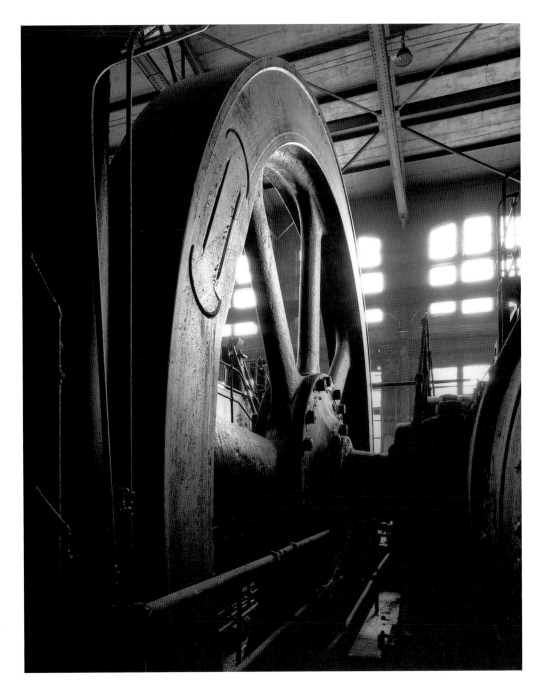

Flywheel of a blowing engine.

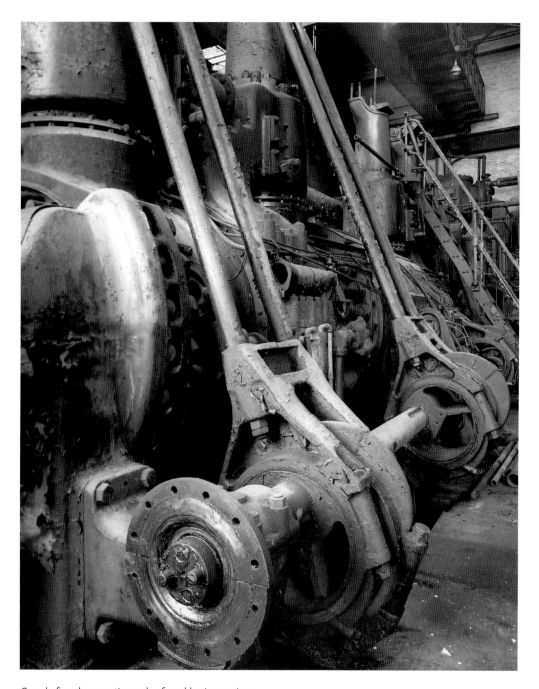

Camshaft and connecting rods of gas-blowing engines.

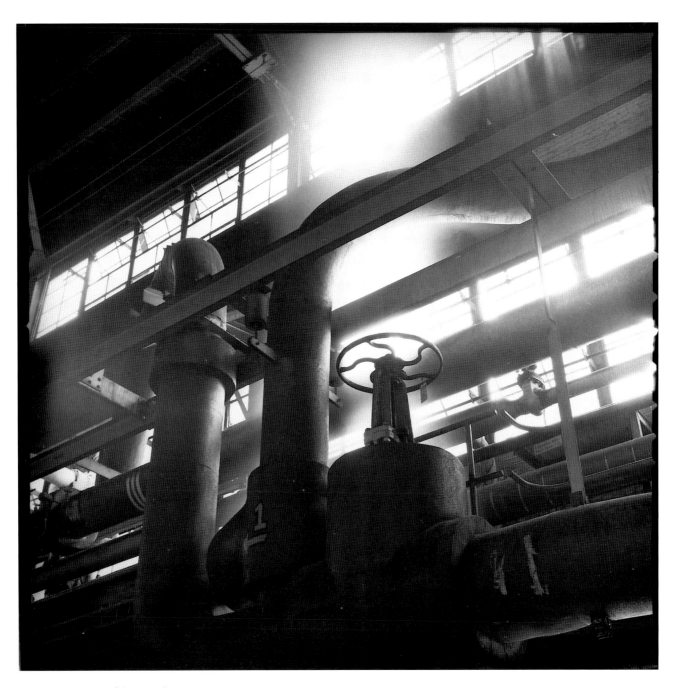

Steam piping in one of the power houses.

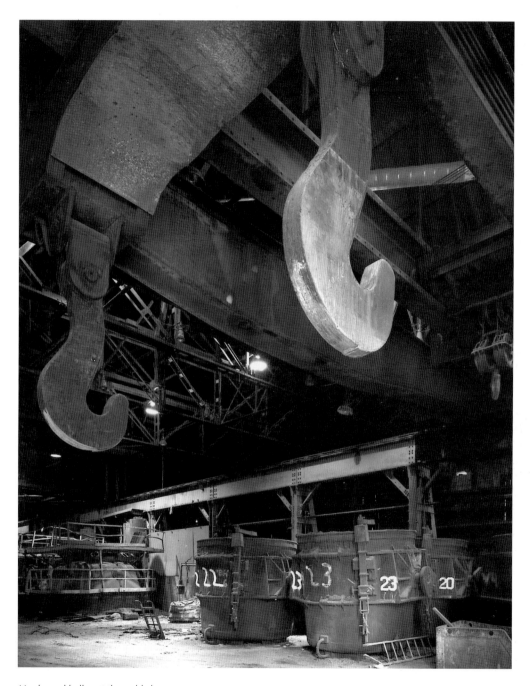

Hooks and ladles at the weld shop.

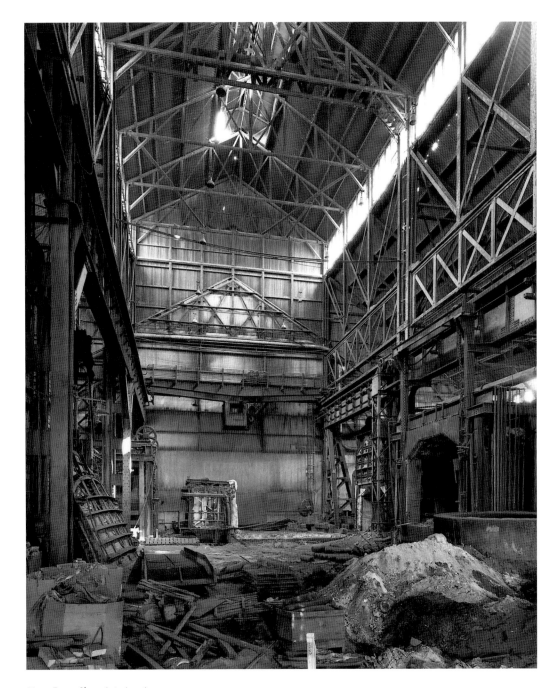

No. 1 Forge Shop, interior view.

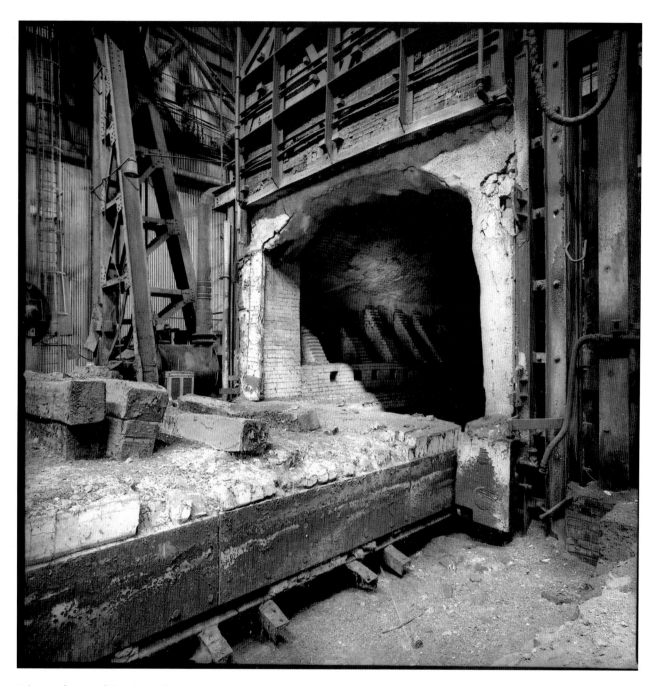

Reheating furnace of No. 1 Forge Shop.

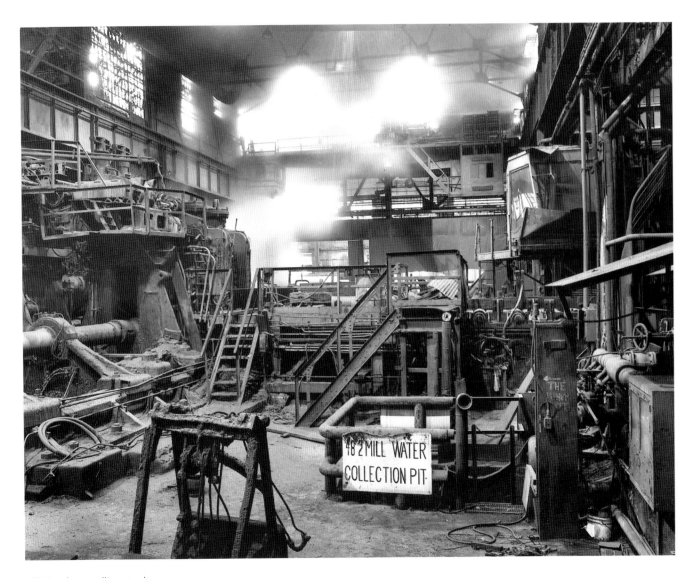

48" Grey beam rolling stand.

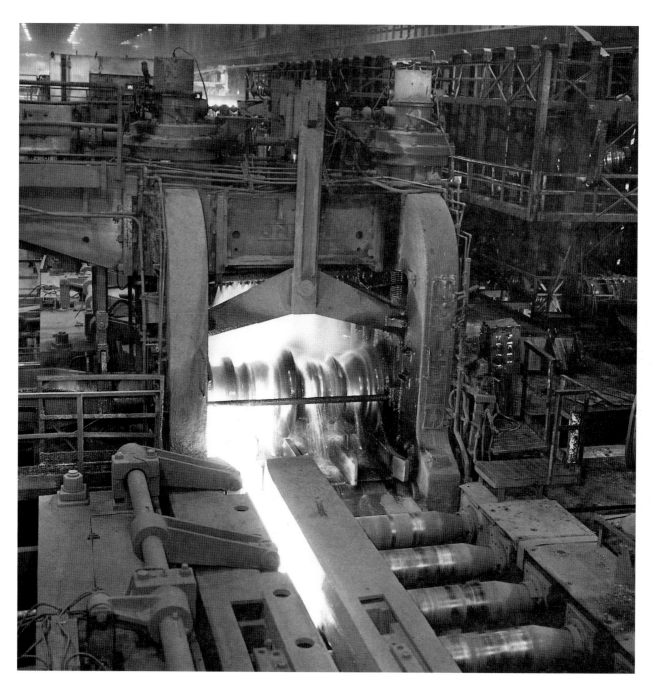

Molten steel beam passing through the combination mill.

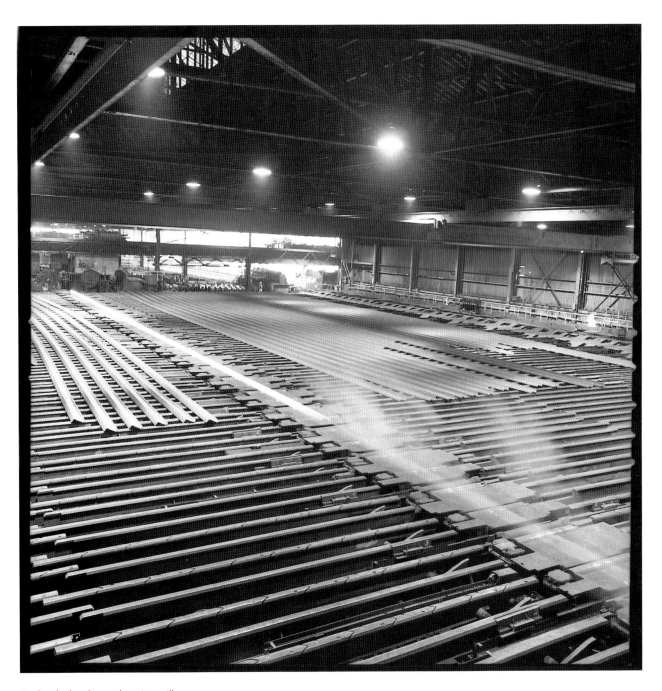

Cooling bed at the combination mill.

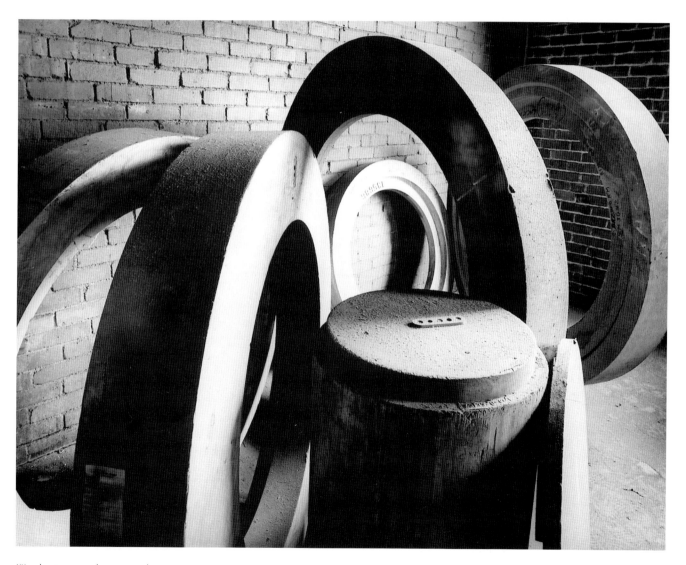

Wood patterns at the pattern shop.

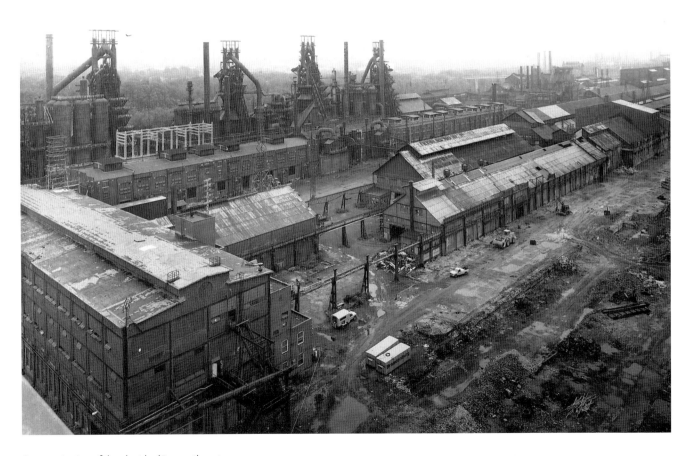

Panoramic view of the plant looking northwest.

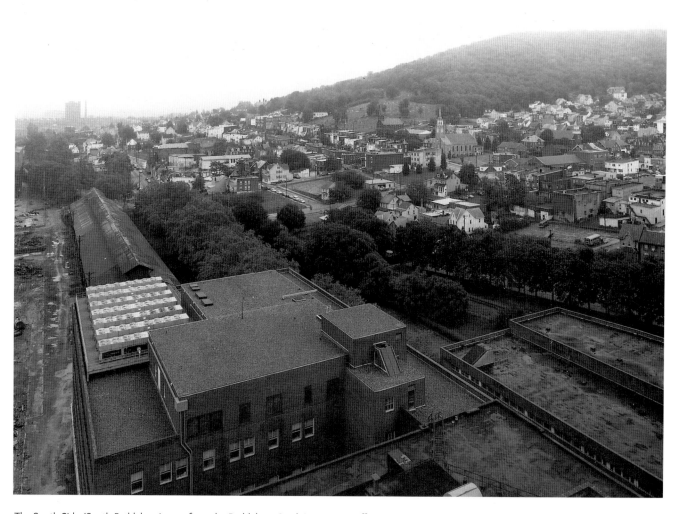

The South Side (South Bethlehem) seen from the Bethlehem Steel Corporation office.

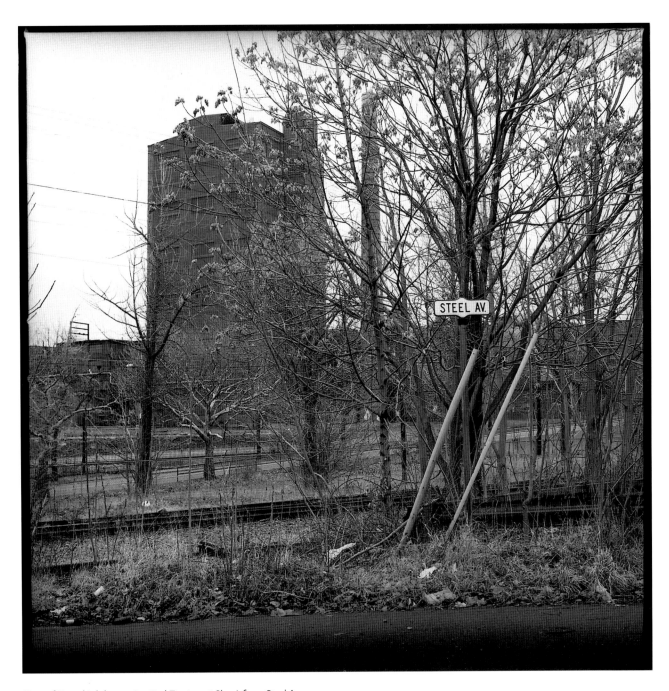

View of No. 5 high house (vertical Treatment Shop) from Steel Avenue.

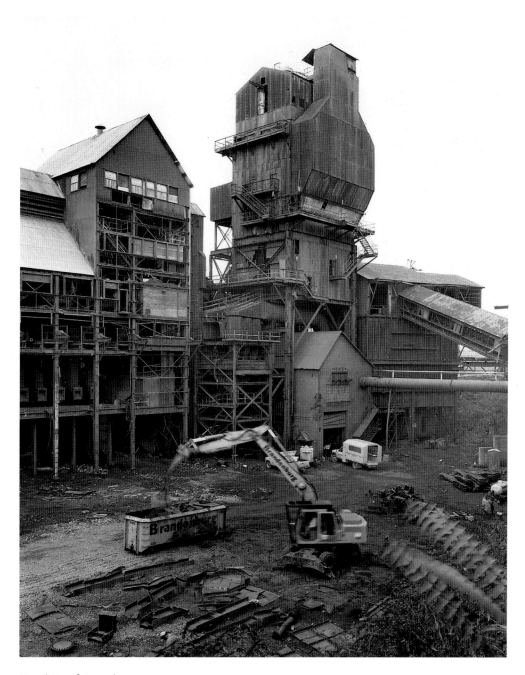

Demolition of sinter plant, 1997.

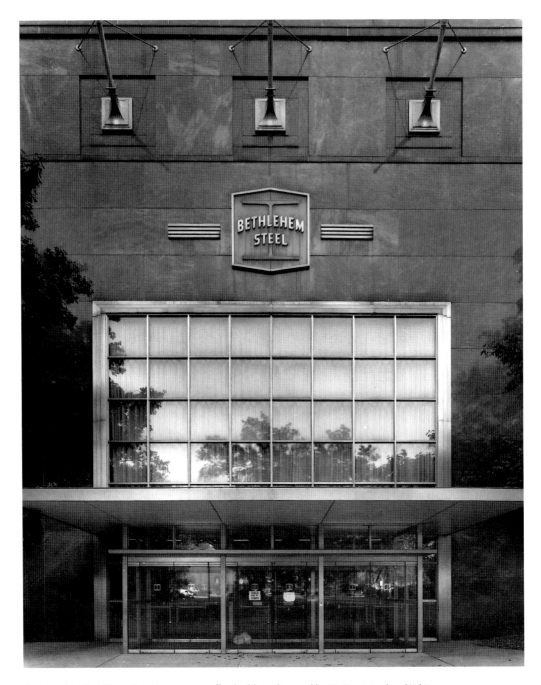

Façade of the Bethlehem Steel Corporation office building, designed by McKim, Mead and White.

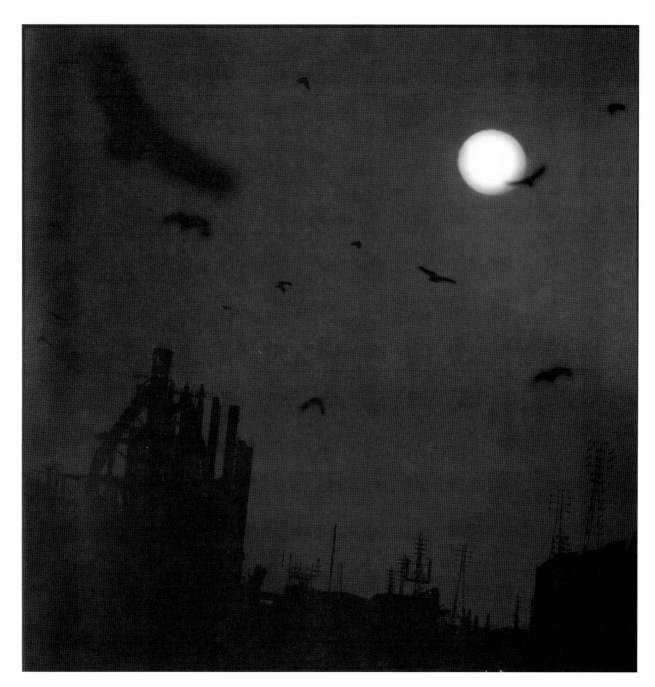

Blast furnaces at night with crows.

Photo Credits